Order Form — USD price for USA only

CW00369455

ISCN 2009

**An International System for
Human Cytogenetic Nomenclature (2009)**
ISBN 3–8055–8985–7

Payment
○ Check enclosed
○ Please bill me
○ Please charge thi
   ○ American Exp
   ○ Eurocard

Card No.:

Exp. date:

Name/address:

Date/Signature

KARGER

Order more – pay less!

Please send:
○ 1 copy: USD 43.00
○ 2 copies: USD 41.00 each
○ 3 copies: USD 40.00 each
○ 5 copies: USD 38.00 each
○ 10 copies: USD 34.00 each
○ 20 copies: USD 32.00 each

○ Five-copy sets of fold-out available of
**The Normal Human Karyotype,
G- and R- bands**
reproduced from ISCN 2005
USD 27.00 per set
ISBN 3–8055—8046–0

Postage and handling free with prepayment

Order Form — CHF and EUR* prices for outside USA

ISCN 2009

**An International System for
Human Cytogenetic Nomenclature (2009)**
ISBN 3–8055—8985–7

Payment
○ Check enclosed
○ Please bill me
○ Please charge this order to my credit card
   ○ American Express   ○ Visa   ○ Diners
   ○ Eurocard   ○ Mastercard

Card No.:

Exp. date:

Name/address:

Date/Signature

KARGER

Order more – pay less!

Please send:
○ 1 copy: CHF 43.–/EUR 30.50
○ 2 copies: CHF 41.–/EUR 29.50 each
○ 3 copies: CHF 40.–/EUR 28.50 each
○ 5 copies: CHF 38.–/EUR 27.– each
○ 10 copies: CHF 34.–/EUR 24.50 each
○ 20 copies: CHF 32.–/EUR 23.– each

○ Five-copy sets of fold-out available of
**The Normal Human Karyotype,
G- and R- bands**
reproduced from ISCN 2005
CHF 30.—/EUR 21.50 per set
ISBN 3—8055–8046–0

Postage and handling free with prepayment
* EUR prices for Germany only

Published in
collaboration with   **Cytogenetic and
Genome Research**

Published in collaboration with *Cytogenetic and Genome Research* under the title *ISCN 2009: An International System for Human Cytogenetic Nomenclature (2009)*
VI + 138 p. plus fold-out, 11 fig., 4 tab., 2009

Although produced by the editors and publishers of *Cytogenetic and Genome Research*, this publication is independent of the subscription.

Copies of ISCN 2009 can be ordered from the publishers:

S. Karger AG
Allschwilerstrasse 10
P.O. Box
CH–4009 Basel
Switzerland
Tel. +41 61 306 11 11
Fax +41 61 306 12 34
E-Mail karger@karger.ch

S. Karger Publishers, Inc.
26 West Avon Road
P.O. Box 529
Unionville, CT 06085
Tel. (860) 675-7834
Fax (860) 675-7302
Toll free: 1-800-828-5479

Library of Congress Cataloging-in-Publication Data

International Standing Committee on Human Cytogenetic Nomenclature.
   ISCN 2009 : an international system for human cytogenetic nomenclature
(2009) / editors, Lisa G. Shaffer, Marilyn L. Slovak, Lynda J. Campbell.
     p. ; cm.
   Includes bibliographical references and index.
   ISBN 978-3-8055-8985-7 (pbk. : alk. paper)
   1.   Human cytogenetics–Nomenclature.   2.   Human cytogenetics–Terminology.
3.   Human chromosomes–Nomenclature.   4.   Human chromosomes–Terminology.
5.   Human chromosome abnormalities–Nomenclature.   6.   Human chromosome
abnormalities–Terminology.   I. Shaffer, Lisa G.   II. Slovak, Marilyn L.   III. Campbell,
Lynda J.   IV. Cytogenetic and genome research.   V. Title.   VI. Title: International
system for human cytogenetic nomenclature (2009).
   [DNLM: 1.   Cytogenetics–Terminology–English.   QU 15 I605i 2009]
   QH431.I56 2009
   611'.01816–dc22
                                                                    2009001276

S. Karger · Medical and Scientific Publishers
Basel · Freiburg · Paris · London · New York · Bangalore ·
Bangkok · Shanghai · Singapore · Tokyo · Sydney

ISBN 978–3–8055–8985–7

Table of Contents

1 Historical Introduction

1.1 1956–1984[1]

In 1956 Tjio and Levan, in their now classic article, reported that the human chromosome number was 46 and not 48. This work, which was carried out on cultured human embryonic cells, was rapidly confirmed by studies of testicular material by Ford and Hamerton (1956). These two articles stimulated a renewed interest in human cytogenetics, and, by 1959, several laboratories were engaged in the study of human chromosomes and a variety of classification and nomenclature systems had been proposed. This resulted in confusion in the literature and a need to establish a common system of nomenclature that would improve communication between workers in the field.

For this reason, a small study group was convened in Denver, Colorado, at the suggestion of Charles E. Ford. Fourteen investigators and three consultants participated, representing each of the laboratories that had published human karyotypes up to that time. The system proposed in the report of this meeting, entitled "A Proposed Standard System of Nomenclature of Human Mitotic Chromosomes," more commonly known as the Denver Conference (1960), has formed the basis for all subsequent nomenclature reports and has remained virtually unaltered, despite the rapid developments of the last 25 years. It is fair to say that the participants at Denver did their job so well that this report has formed the cornerstone of human cytogenetics since 1960, and the foresight and cooperation shown by these investigators have prevented much of the nomenclature confusion which has marked other areas of human genetics.

Three years later, a meeting called by Lionel S. Penrose was held in London (London Conference, 1963) to consider developments since the Denver Conference. The most significant result of that conference was to give official sanction to the classification of the seven groups of chromosomes by the letters A to G, as originally proposed by Patau (1960).

The next significant development came in Chicago at the Third International Congress on Human Genetics in 1966 when 37 investigators, representing the major cytogenetic laboratories, met to determine whether it was possible to improve the nomenclature and thus eliminate some of the major problems that had resulted from the rapid proliferation of new findings since 1960. The report of this conference (Chicago Conference, 1966) proposed a standard system of nomenclature for the provision of short-hand descriptions of the human chromosome complement and its abnormalities, a system that, in its basic form, has stood the test of time and is now used throughout the world for the description of non-banded chromosomes.

---

[1] Adapted from ISCN (1985).

1

In his introductory address to the Chicago Conference (1966), Lionel Penrose made the following prophetic statement:

"It is easy to be carried away by the detectable peculiarities and to forget that much underlying variability is still hidden from view until some new technical device discloses the finer structure of chromosomes, as in the Drosophila salivary gland cells."

Two years later, in 1968, the second major breakthrough occurred when Torbjörn Caspersson and his colleagues, working in Sweden, published the first banding pictures of plant chromosomes stained with quinacrine dihydrochloride or quinacrine mustard (Caspersson et al., 1968). These studies were rapidly expanded to human chromosomes by these workers, who published the first banded human karyotype in 1970 (for a review of this work, see Caspersson et al., 1972). Soon, several other techniques that also produced chromosome bands were developed. This led to the realization that, as each human chromosome could now be identified very precisely, the existing system of nomenclature would no longer be adequate.

A group of 50 workers concerned with human cytogenetics met in 1971 on the occasion of the Fourth International Congress of Human Genetics in Paris to agree upon a uniform system of human chromosome identification. Their objective was accomplished and extended by the appointment of a Standing Committee, chaired by John Hamerton, which met initially in Edinburgh in January 1972, and then with a number of expert consultants at Lake Placid in New York in December 1974, and again in Edinburgh in April 1975.

The 1971 meeting in Paris, together with the 1972 Edinburgh meeting of the Standing Committee, resulted in the report of the Paris Conference (1971), a highly significant document in the annals of human cytogenetics. This document proposed the basic system for designating not only individual chromosomes but also chromosome regions and bands, and it provided a way in which structural rearrangements and variants could be described in terms of their band composition.

By 1974 it had become clear that the number of workers in the field was now too great to allow the holding of such conferences as the Chicago and Paris ones, where the majority of laboratories involved could be represented. The Standing Committee therefore proposed holding smaller, nonrepresentative conferences, each on a number of fairly specific topics and that would utilize expert consultants for each topic. The first meeting of this type was held in 1974 in Lake Placid and the second in 1975 in Edinburgh, at which a number of specific topics – including heteromorphic chromosomes of the Hominoidea, and chromosome registers – were discussed. These discussions were reported in the 1975 supplement to the Paris Conference report (Paris Conference, 1971, Supplement, 1975).

A further change came about in 1976 at the Fifth International Congress of Human Genetics in Mexico City, when a meeting of all interested human cytogeneticists was held to elect an International Standing Committee on Human Cytogenetic Nomenclature. These elections provided a truly international and geographic representation for the Standing Committee and provided a mandate to the committee to continue its work in proposing ways in which human chromosome nomenclature might be improved. Jan Lindsten was appointed the chairman of this committee.

The committee met in Stockholm in 1977 and, following past practice, invited a number of expert consultants to meet with it. It was decided at this meeting to cease labeling reports geographically and to unify the various conference reports reviewed above into a document entitled "An International System for Human Cytogenetic Nomenclature (1978)," to be abbreviated ISCN (1978). ISCN (1978) included all major decisions of the Denver, London, Chicago, and Paris Conferences, without any major changes but edited for consistency and accuracy. It thus provided in one document a complete system of human cytogenetic nomenclature that has stood the test of time and has proved to be of value not only to those entering the field for the first time but also to experienced cytogeneticists.

The next major area to be considered by the Standing Committee was the nomenclature of chromosomes stained to show "high resolution banding." In 1977 a working party was established under the chairmanship of Bernard Dutrillaux to consider this matter.

It had been recognised for some time that prophase and prometaphase chromosomes reveal a much larger number of bands than can be seen even in the best banded metaphase chromosome preparations. Techniques were devised to partially synchronise peripheral blood cultures so as to yield sufficient cells in the early phase of mitosis for detailed study. These all essentially use some method of blocking cells in the S-phase, releasing the block and then timing the subsequent harvest to obtain the maximum number of cells at the appropriate stage (Dutrillaux, 1975; Yunis, 1976). Several studies showed that techniques of this kind required a new nomenclature (Francke and Oliver, 1978; Viegas-Pequignot and Dutrillaux, 1978; Yunis et al., 1978).

The working group met on several occasions. There was a remarkable degree of agreement on the number of bands, the width of the bands and their relative positions. There was, however, considerable difficulty in reaching a consensus on the origin of certain bands and on the stage of their appearance relative to other bands. A broad measure of agreement was, however, reached at a meeting in Paris in May 1980 and this was published as "An International System for Human Cytogenetic Nomenclature – High Resolution Banding (1981)" or ISCN (1981).

A new Standing Committee was elected at a specially convened meeting of cytogeneticists held during the Sixth International Congress of Human Genetics in Jerusalem in 1981. David Harnden was appointed chairman of the new committee.

A revision of the International System for Human Cytogenetic Nomenclature was prepared in 1984, to be published as ISCN (1985), partly because a reprint was in any case necessary and partly because, once again, it was felt to be important to try to keep all statements on nomenclature together in a single volume. The opportunity was taken to correct errors and make a small number of amendments but no attempt was made to make a major revision.

The widely accepted international nomenclature for human chromosomes has proved to be an important element in improving and maintaining international collaboration. The development of this system has been made possible by the collaboration of many people. I would like to thank not only members of the Standing Committee but others who have acted as consultants or who have contributed ideas or materials to these publications. In particular I would like to express the gratitude of the international cytogenetic community to the March of Dimes Birth Defects Foundation for its consistent and substantial financial support over the past 19 years. Without its help none of these developments would have been possible.

David Harnden
October 1984

1.2   1985–1995

A new Standing Committee was elected at a meeting of cytogeneticists attending the Seventh Congress of Human Genetics held in Berlin in 1986 and Uta Francke was appointed as chairman. The Committee was aware of a considerable increase in the amount and variety of data on chromosome aberrations associated with neoplasia, and considered that a terminology was necessary for those acquired chromosome aberrations that were not adequately described by the nomenclature for constitutional aberrations as published in ISCN (1985). A subcommittee under the chairmanship of Felix Mitelman was established and charged with the task of producing a nomenclature for cancer cytogenetics. The report of this subcommittee was adopted by the ISCN Standing Committee and published as "ISCN (1991): Guidelines for Cancer Cytogenetics". These guidelines superseded previous ISCN recommendations on cancer cytogenetics and have since come into general use.

A new Standing Committee was elected at the Eighth International Congress of Human Genetics held in Washington, DC, in 1991 and Felix Mitelman was appointed chairman. The new committee considered that it would be timely to review and update the ISCN (1985) nomenclature in the light of developments in the field, including advances in the use of in situ hybridization techniques, and to incorporate all revisions and the guidelines for cancer cytogenetics into a single document to be published as ISCN (1995). Cytogeneticists were asked, through notices published in relevant journals, to forward to the Committee their comments on any defects in the ISCN 1978–1991 publications, as well as any suggestions for alterations and improvements.

The Standing Committee and consultants met in Memphis on October 9–13, 1994, at the kind invitation of Professor Avirachan Tharapel. The Committee considered all the recommendations that had been submitted to it and updated, modified and amalgamated the 1985 and 1991 documents into a single text with the intention of this being published in 1995.

H.J. Evans
P.A. Jacobs
October 1994

1.3   1996–2004

The Ninth International Congress of Human Genetics was held in Rio de Janeiro in 1996. A new Standing Committee was elected at the satellite meeting of the cytogeneticists. Patricia A. Jacobs was appointed as the chairperson of the Committee. In light of the extensive revision of ISCN (1995), the new Committee elected not to implement additional changes during its term.

The Tenth International Congress of Human Genetics was held in Vienna. The congregation of cytogeneticists present at the satellite meeting elected a new committee and Niels Tommerup was appointed as chairman. The extensive use of ISCN (1995) by the scientific community identified several areas that needed clarifications, deletions and additions. Therefore, the Committee decided to review and update the ISCN (1995). The seven members of the Committee and nine external consultants met in Vancouver, BC, December 8–10, 2004 at the invitation of Niels Tommerup and Lisa G. Shaffer. The primary changes included replacing G- and R-banded karyotypes (Figs. 2 and 3) with new ones reflecting higher band-level resolutions, the addition of a new idiogram at the 300-band level and introduction of a new 700-band level idiogram that reflected the actual size and position of bands. The in situ hybridization nomenclature was modernized, simplified, and expanded. New examples reflecting unique situations were added, and a basic nomenclature for recording array comparative genomic hybridization results was introduced.

The Committee adopted changes to its membership structure for the future. The number of members was expanded to eleven from the current seven to reflect better representation of the geographic distribution of cytogeneticists. The voting constituency and guidelines for the election of members and chairpersons was redefined. Lisa Shaffer was appointed as chairman of the newly elected Committee. Finally, the Committee recommended that ISCN (2005) be published in 2005.

D.H. Ledbetter
A.T. Tharapel
December 2004

1.4   2005–2009

Early in 2006, Lisa Shaffer and Niels Tommerup organized the election for the next Standing Committee. Ballots were distributed and collected worldwide, and at the Eleventh International Congress of Human Genetics, held in Brisbane, Australia, in 2006, the results of the election were announced, resulting in a Committee of eleven elected members. The newly elected Committee received feedback on ISCN (2005) and decided to hold a meeting in 2008 to discuss potential changes and additions to a new edition of ISCN. At the invitation of Lisa Shaffer, chair, the Committee and two external consultants met in Vancouver, BC, October 8–10, 2008. The primary change in cancer was the accommodation for either **idem** or **sl/sdl** in the nomenclature to describe clonal evolution. The in situ hybridization nomenclature was further clarified and additional examples provided. The basic microarray nomenclature was revised and expanded to accommodate all platform types, with more examples provided. Finally, a nomenclature for MLPA was introduced. The Committee recommended that ISCN (2009) be published in 2009.

Lisa G. Shaffer
Marilyn L. Slovak
Lynda J. Campbell
December 2008

2   Normal Chromosomes

2.1   Introduction

Human chromosome nomenclature is based on the results of several international conferences (Denver 1960, London 1963, Chicago 1966, Paris 1971, Paris 1975, Stockholm 1977, Paris 1980, Memphis 1994, Vancouver 2004). The present report, which summarizes the current nomenclature, incorporates and supersedes all previous ISCN recommendations. The ISCN Standing Committee recommends that this nomenclature system be used also in other species.

2.2   Chromosome Number and Morphology

2.2.1   Non-Banding Techniques

In the construction of the karyogram[1] the autosomes are numbered from 1 to 22 in order of decreasing length (one exception is that chromosome 21 is shorter than chromosome 22). The sex chromosomes are referred to as X and Y.

When the chromosomes are stained by methods that do not produce bands, they can be arranged into seven readily distinguishable groups (A–G) based on descending order of size and the position of the centromere.

The group letter designations placed before the chromosome numbers are those agreed upon at the London Conference (1963). Not all chromosomes in the D and G groups show satellites on their short arms in a single cell. The number and size of these structures are variable.

The following parameters were used to describe non-banded chromosomes: (1) the length of each chromosome, expressed as a percentage of the total length of a normal haploid set, i.e., the sum of the lengths of the 22 autosomes and of the X chromosome; (2) the arm ratio of the chromosomes, expressed as the length of the longer arm relative to the shorter one; and (3) the centromeric index, expressed as the ratio of the length of the shorter arm to the whole length of the chromosome. The latter two indices are, of course, related algebraically.

---

[1]   The terms *karyogram, karyotype* and *idiogram* have often been used indiscriminately. The term *karyogram* should be applied to a systematized array of the chromosomes prepared either by drawing, digitized imaging, or by photography, with the extension in meaning that the chromosomes of a single cell can typify the chromosomes of an individual or even a species. The term *karyotype* should be used to describe the normal or abnormal, constitutional or acquired, chromosomal complement of an individual, tissue or cell line. We recommend that the term *idiogram* be reserved for the diagrammatic representation of a karyotype.

Group A (1–3)	Large metacentric chromosomes readily distinguished from each other by size and centromere position.
Group B (4–5)	Large submetacentric chromosomes.
Group C (6–12, X)	Medium-sized metacentric or submetacentric chromosomes. The X chromosome resembles the longer chromosomes in this group.
Group D (13–15)	Medium-sized acrocentric chromosomes with satellites.
Group E (16–18)	Relatively short metacentric or submetacentric chromosomes.
Group F (19–20)	Short metacentric chromosomes.
Group G (21–22, Y)	Short acrocentric chromosomes with satellites. The Y chromosome bears no satellites.

2.2.2   Banding Techniques

Numerous technical procedures have been reported that produce banding patterns on metaphase chromosomes.

A **band** is defined as the part of a chromosome that is clearly distinguishable from its adjacent segments by appearing darker or lighter with one or more banding techniques. Bands that stain darkly with one method may stain lightly with other methods. The chromosomes are visualized as consisting of a continuous series of light and dark bands, so that, by definition, there are no "interbands".

The methods first published for demonstrating bands along the chromosomes were those that used quinacrine mustard or quinacrine dihydrochloride to produce a fluorescent banding pattern. These methods are named Q-staining methods and the resulting bands **Q-bands** (Fig. 1). The numbers assigned to each chromosome were based on the Q-banding pattern as given by Caspersson et al. (1972). Techniques that demonstrate an almost identical pattern of dark and light bands along the chromosomes usually use the Giemsa dye mixture as the staining agent. These techniques are generally termed G-staining methods and the resulting bands **G-bands** (Fig. 2). Some banding techniques give patterns that are opposite in staining intensity to those obtained by the G-staining methods, viz, the reverse staining methods, and the resulting bands are called **R-bands** (Fig. 3).

The banding techniques fall into two principle groups: (1) those resulting in bands distributed along the length of the whole chromosome, such as G-, Q-, and R-bands, including techniques that demonstrate patterns of DNA replication, and (2) those that stain specific chromosome structures and hence give rise to a restricted number of bands (Table 1). These include methods that reveal constitutive heterochromatin (**C-bands**) (Fig. 4), telomeric bands (**T-bands**), and nucleolus organizing regions (**NORs**). For the code to describe banding techniques, see Table 2.

The patterns obtained with the various C-banding methods do not permit identification of every chromosome in the somatic cell complement but, as demonstrated in Table 1, can be used to identify specific chromosomes. The C-bands on chromosomes 1, 9, 16, and Y are all morphologically variable. The short-arm regions of the acrocentric chromosomes also demonstrate variations in size and staining intensity of the Q-, G-, R-, C-, T-, and NOR-bands. These variations are heritable features of the particular chromosome.

Normal Chromosomes

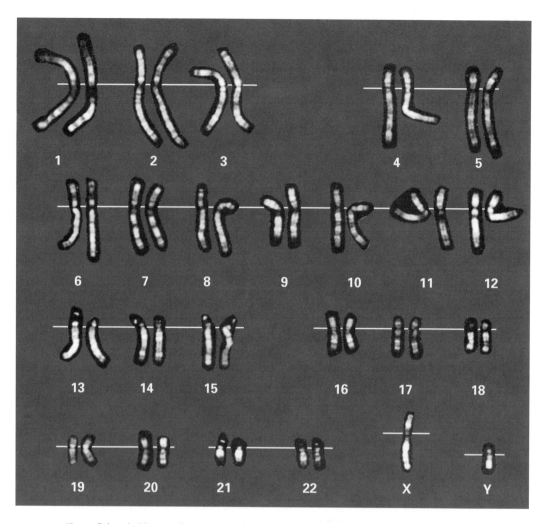

Fig. 1. Q-banded human karyogram. (Courtesy of Dr. E. Magenis.)

2.2.3   X- and Y-Chromatin

Inactive X chromosomes, as well as the heterochromatic segment on the long arm of the Y chromosome, appear as distinctive structures in interphase nuclei, for which the terms **X-chromatin** (Barr body, sex chromatin, X-body) and **Y-chromatin** (Y-body), respectively, should be used.

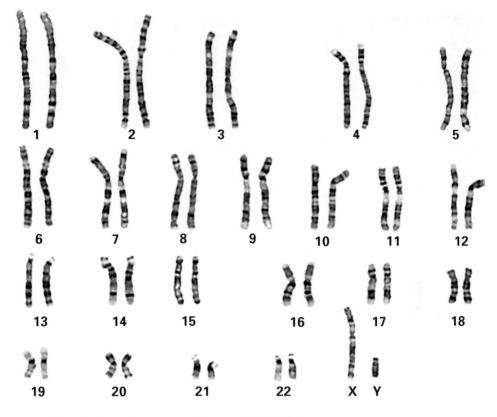

Fig. 2. G-banded human karyogram. (Courtesy of N.L. Chia.)

2.3 Chromosome Band Nomenclature

2.3.1 Identification and Definition of Chromosome Landmarks, Regions, and Bands

Each chromosome in the human somatic cell complement is considered to consist of a continuous series of bands, with no unbanded areas. As defined earlier, a **band** is a part of a chromosome clearly distinguishable from adjacent parts by virtue of its lighter or darker staining intensity. The bands are allocated to various regions along the chromosome arms, and the regions are delimited by specific **landmarks**. These are defined as consistent and distinct morphologic features important in identifying chromosomes. Landmarks include the ends of the chromosome arms, the centromere, and certain bands. The bands and the regions are numbered from the centromere outward. A **region** is defined as an area of a chromosome lying between two adjacent landmarks.

The original banding pattern was described in the Paris Conference (1971) report and was based on the patterns observed in different cells stained with either the Q-, G-, or R-banding technique (Appendix, Chapter 17). The banding patterns obtained

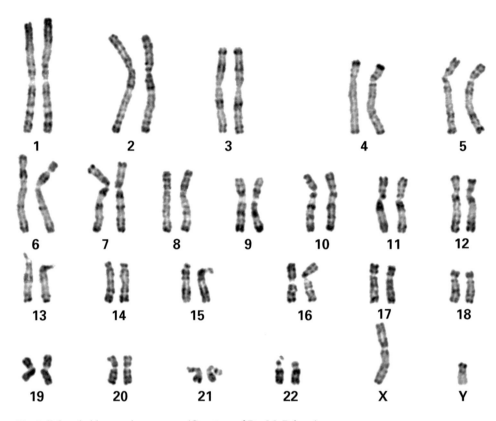

Fig. 3. R-banded human karyogram. (Courtesy of Dr. M. Prieur.)

with these staining methods agreed sufficiently to allow the construction of a single diagram representative of all three techniques. The bands were designated on the basis of their midpoints and not by their margins. Intensity was taken into consideration in determining which bands should serve as landmarks on each chromosome in order to divide the chromosome into natural, easily recognizable morphologic regions. A list of bands serving as landmarks which were used in constructing this diagram is provided in Table 3.

2.3.2 Designation of Regions, Bands, and Sub-Bands

Regions and bands are numbered consecutively from the centromere outward along each chromosome arm. The symbols **p** and **q** are used to designate, respectively, the short and long arms of each chromosome. The centromere (**cen**) itself is designated 10; the part facing the short arm is p10, the part facing the long arm is q10. These are not shown in the idiograms. The two regions adjacent to the centromere are labeled as 1 in each arm; the next, more distal regions as 2, and so on. A band used as a landmark is considered as belonging entirely to the region distal to the landmark and is accorded the band number of 1 in that region.

Table 1. Examples of heteromorphisms with various stains[a]

Technique	Chromosome									
	1	2	3	4	5	6	7	8	9	10
G[b]	q12 inv(p13q21)	inv(p11.2q13)	inv(p11.2q12)			p11.1			q12 inv(p12q13)	inv(p11.2q21.2)
C[c]	qh								qh	
G11	qh		cen		q11.1		p11.1		qh	q11.1
R or T	p36.3	q37		p16	p15.3 q35		p22	q24.3	q34	q26
NOR										
Q[d]			cen	cen						
DA-DAPI[e]	qh								qh	

Technique	Chromosome													
	11	12	13	14	15	16	17	18	19	20	21	22	X	Y
G[b]			p	p	p	q11.2 inv(p11.2q12.1)					p	p		inv(p11.2q11.2)
C[c]			p	p	p	qh	p11				p	p		q12
G11			p	p	p		p11.1			q11.1	p	p		q12
R or T	p15 q13	p13	p12 q34	p12 q32	p12	p13.3 q24	q25		p13.3 q13.1	q13	p12 q22	p12 q11.2 q13		
NOR			p12	p12	p12						p12	p12		
Q[d]			p11.2 p13 cen	p11.2 p13	p11.2 p13						p11.2 p13	p11.2 p13		
DA-DAPI[e]					p11.2	qh								q12

[a] cen = centromere, h = heterochromatin, inv = inversion, p = short arm, q = long arm.
[b] Only the most commonly seen heteromorphisms are listed.
[c] All centromeres show constitutive heterochromatin variation.
[d] Only the brilliant and intensity-variable Q-bands are listed.
[e] DA-DAPI = Distamycin A and 4′,6-diamidino-2-phenylindole.

In designating a particular band, four items are required: (1) the chromosome number, (2) the arm symbol, (3) the region number, and (4) the band number within that region. These items are given in order without spacing or punctuation. For example, 1p31 indicates chromosome 1, short arm, region 3, band 1.

Whenever an existing band is subdivided, a decimal point is placed after the original band designation and is followed by the number assigned to each sub-band. The sub-bands are numbered sequentially from the centromere outward. For example, if the original band 1p31 is subdivided into three equal or unequal sub-bands, the sub-bands are labeled 1p31.1, 1p31.2, and 1p31.3, sub-band 1p31.1 being proximal and 1p31.3 distal to the centromere. If a sub-band is subdivided, additional digits, but no further punctuation, are used; e.g., sub-band 1p31.1 might be further subdivided into 1p31.11, 1p31.12, etc. Although in principle a band can be subdivided into any number of new bands at any one stage, a band is usually subdivided into three sub-bands.

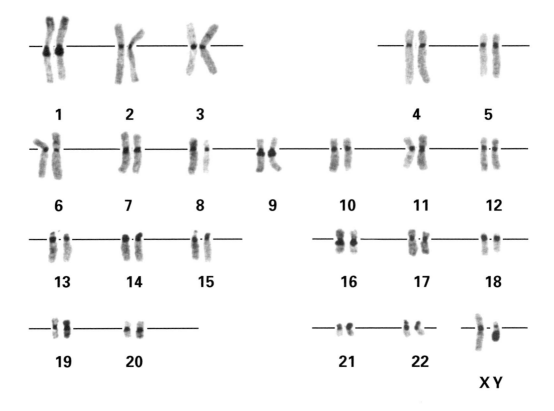

Fig. 4. C-banded human karyogram. (Courtesy of Dr. N. Mandahl.)

2.4 High-Resolution Banding

The nomenclature for high-resolution preparations of prophase and prometaphase chromosomes set forth by ISCN (1981) is an extension of the nomenclature for the banding patterns for metaphase chromosomes established at the Paris Conference (1971) and in ISCN (1978). The original system was specifically devised to allow for expansion as more chromosome bands were recognized.

High-resolution banding techniques can be applied to chromosomes in different stages of the cell cycle, e.g., prophase, prometaphase or interphase (by methods that induce premature chromosome condensation). Furthermore, the number of discernible bands depends not only on the state of condensation but also on the banding technique used. The level of resolution is determined by the number of bands seen in a haploid set (22 autosomes + X and Y). The standard idiograms shown in Fig. 5 provide schematic representations of chromosomes corresponding to approximately 300, 400, 550, 700 and 850 bands. Although larger numbers of bands can be visualized, 550- to 850-band idiograms are sufficient for practical purposes. The 400- and 550-band idiograms are taken from ISCN (1985), and the 850-band idiogram was introduced in ISCN (1981) (Francke, 1994). The original nomenclature was

Table 2. Examples of the code used to describe banding techniques. In this one-, two-, or three-letter code, the first letter denotes the type of banding, the second letter the general technique, and the third letter the stain.

Q	Q-bands
QF	Q-bands by fluorescence
QFQ	Q-bands by fluorescence using quinacrine
QFH	Q-bands by fluorescence using Hoechst 33258
G	G-bands
GT	G-bands by trypsin
GTG	G-bands by trypsin using Giemsa
GTL	G-bands by trypsin using Leishman
GTW	G-bands by trypsin using Wright
GAG	G-bands by acetic saline using Giemsa
C	C-bands
CB	C-bands by barium hydroxide
CBG	C-bands by barium hydroxide using Giemsa
R	R-bands
RF	R-bands by fluorescence
RFA	R-bands by fluorescence using acridine orange
RH	R-bands by heating
RHG	R-bands by heating using Giemsa
RB	R-bands by BrdU
RBG	R-bands by BrdU using Giemsa
RBA	R-bands by BrdU using acridine orange
DA-DAPI	DAPI-bands by Distamycin A and 4',6-diamidino-2-phenylindole

based on patterns rather than on measurements. Also, variation in intensity of staining, which is dependent on the staining technique, was not reflected in the ISCN (1981) idiograms. At higher resolution, however, an idiogram depicting only patterns of black and white bands becomes difficult to use. Therefore, the ISCN (1981) 850-band idiogram has been replaced by previously published idiograms that are based on measurements of trypsin-Giemsa bands on prometaphase chromosomes and include five different shades of staining intensities to facilitate orientation among the large number of bands (Francke, 1981, 1994).

The ISCN (2005) added the 300- and 700-band idiograms as additional references. The idiograms, which show a G-band pattern, are provided to represent the position of bands in G-, Q- and R-stained preparations. Although the appearance of bands visualized by G-staining may differ from that revealed by R-staining (Fig. 6), the sequence of the bands is the same.

Two kinds of variable regions are indicated by different cross-hatching patterns, one involving the pericentromeric heterochromatin regions on all chromosomes and the other involving the variable regions 1q12, 3q11.2, 9q12, 16q11.2, 19p12, 19q12, Yq12 and the short arms of all acrocentric chromosomes. The representations of these variable regions are not based on measurements. Banded structures can be seen within the variable regions, in particular in 1q12, 9q12 and Yq12, but since they are variable they have not been detailed in the idiograms. Normal chromosome variants are discussed in more detail in Chapter 7.

Table 3. Bands serving as landmarks which divide the chromosomes into cytologically defined regions. The omission of an entire chromosome or chromosome arm indicates that either both arms or the arm in question consists of only one region, delimited by the centromere and the end of the chromosome arm.

Chromosome		Number of regions	Landmarks[a]
Number	Arm		
1	p	3	Proximal band of medium intensity (21), median band of medium intensity (31)
	q	4	Proximal negative band (21) distal to variable region, median intense band (31), distal band of medium intensity (41)
2	p	2	Median negative band (21)
	q	3	Proximal negative band (21), distal negative band (31)
3	p	2	Median negative band (21)
	q	2	Median negative band (21)
4	q	3	Proximal negative band (21), distal negative band (31)
5	q	3	Median band of medium intensity (21), distal negative band (31)
6	p	2	Median negative band (21)
	q	2	Median negative band (21)
7	p	2	Distal band of medium intensity (21)
	q	3	Proximal band of medium intensity (21), median band of medium intensity (31)
8	p	2	Median negative band (21)
	q	2	Median band of medium intensity (21)
9	p	2	Median intense band (21)
	q	3	Median band of medium intensity (21), distal band of medium intensity (31)
10	q	2	Proximal intense band (21)
11	q	2	Median negative band (21)
12	q	2	Median band of medium intensity (21)
13	q	3	Median intense band (21), distal intense band (31)
14	q	3	Proximal intense band (21), distal band of medium intensity (31)
15	q	2	Median intense band (21)
16	q	2	Median band of medium intensity (21)
17	q	2	Proximal negative band (21)
18	q	2	Median negative band (21)
21	q	2	Median intense band (21)
X	p	2	Proximal band of medium intensity (21)
	q	2	Proximal band of medium intensity (21)

[a] The numbers in parentheses are the region and band numbers as shown in Fig. 5.

The lowest band number of 10 is assigned to the centromere (not shown on idiograms). The adjacent heterochromatic regions carry band designations of 11, 11.1 or 11.11 depending on the level of resolution.

One problem in assigning numbers to euchromatic sub-bands is that in G-banded preparations new G-bands appear to arise by subdivision of darkly stained G-bands on less extended chromosomes, while in R-staining preparations the dark R-bands appear to split. These interpretations of band to sub-band relationships would lead to different number assignments. Therefore, in assigning sub-band numbers, arbitrary decisions were made for the purposes of nomenclature only that should not be interpreted as statements about chromosome physiology. Examples of G- and R-banded chromosomes at successive stages of resolution are shown in Fig. 6a and b. In addition, G- and R-banded metaphase chromosomes at approximately the 550-band level and their diagrammatic representation (modified from ISCN 1985) are illustrated in a detachable foldout on the inside of the backcover.

2.5   Molecular Basis of Banding

Chromosome bands reflect the functional organization of the genome that regulates DNA replication, repair, transcription, and genetic recombination. The bands are large structures, each approximately 5 to 10 megabases of DNA that may include hundreds of genes. The molecular basis of banding methods is known to involve nucleotide base composition, associated proteins, and genome functional organization. In general, Giemsa-positive bands (G-dark bands, R-light bands) are AT-rich, late replicating, and gene poor; whereas, Giemsa-negative bands (G-light bands, R-dark bands) are CG-rich, early replicating, and relatively gene rich.

Centromeric DNA and pericentromeric heterochromatin, composed of α-repetitive DNA and various families of repetitive satellite DNA, are easily detected by C-banding. The telomere is composed of 5 to 20 kb of tandem hexanucleotide mini-satellite repeat units, TTAGGG, and stains darkly by T-banding. The 18S and 28S ribosomal RNA genes are clustered together in large arrays containing about 40 copies of each gene. These are located on the acrocentric short arms, at the nucleolar organizer regions or NORs, and are detected by silver staining.

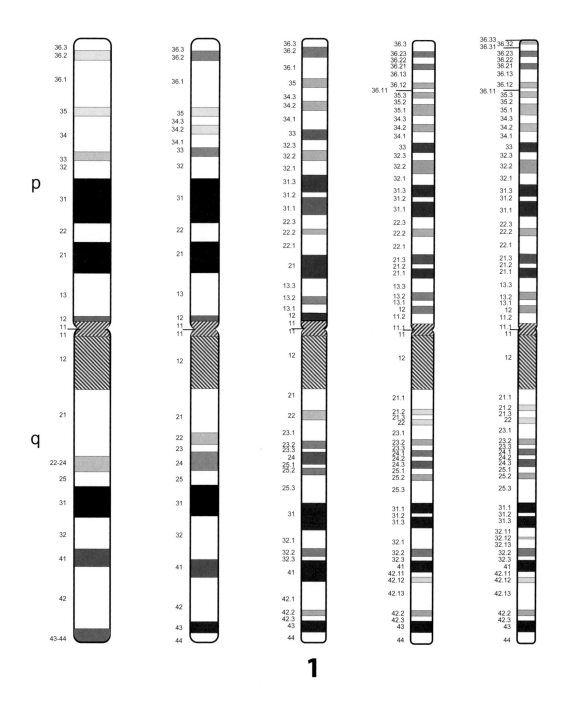

Fig. 5. Idiograms of G-banding patterns for normal human chromosomes at five different levels of resolution. From the left, chromosomes in each group represent a haploid karyotype of approximately 300-, 400-, 550-, 700-, and 850-band levels. The location and width of bands are not based on any measurements. The dark G-bands correspond to bright Q-bands, with the exception of the variable regions. The numbering of R-banded chromosomes is exactly the same, with a reversal of light and dark bands. While the band numbers are exactly the same, the relative widths of euchromatic bands

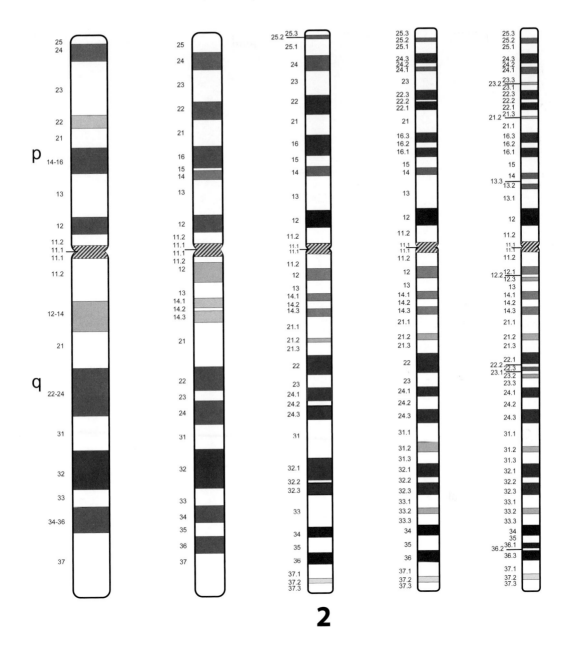

p

q

2

are based on measurements and the staining intensities reflect GTG bands (Francke, 1981, 1994). The idiograms of the Y chromosome are according to observations of Magenis and Barton (1987). While the number of bands on the euchromatic portion of the long arm has been expanded, the designations for light versus dark bands have been maintained. The 400-, 550-, and 850-band idiograms correspond to the ISCN (1995) nomenclature. The 300- and 700-band idiograms, new to ISCN (2005), were provided by N.L. Chia.

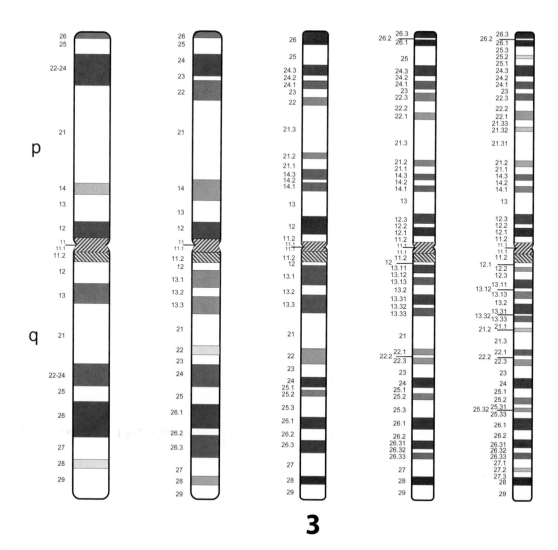

3

ISCN 2009

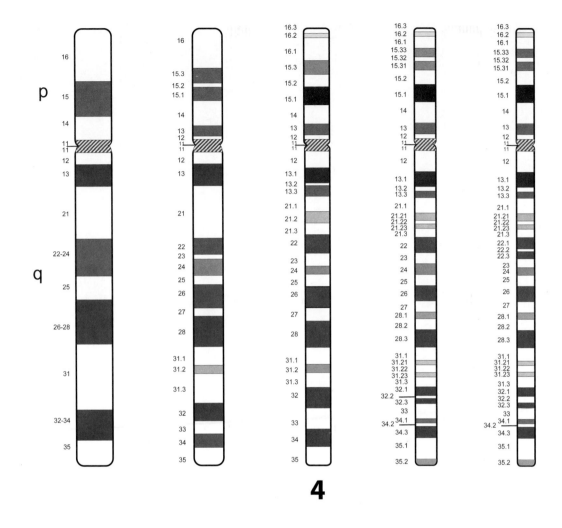

4

Fig. 5 continued (see legend on pp 16–17)

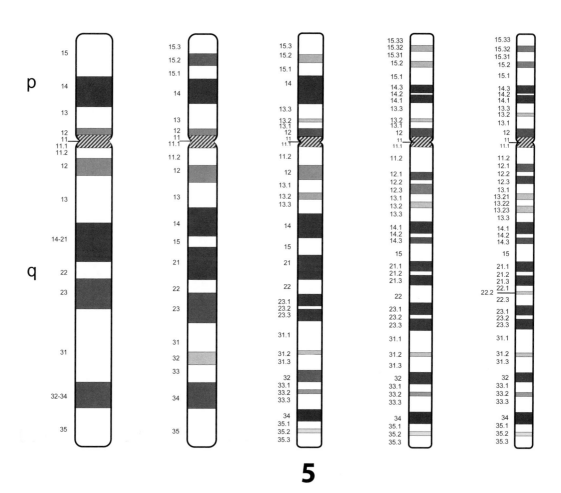

5

ISCN 2009

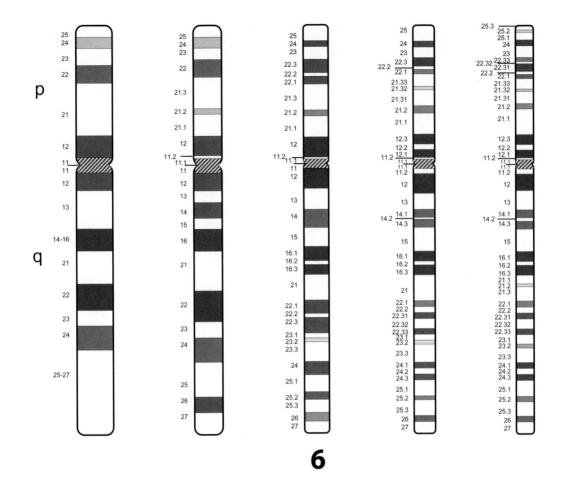

6

Fig. 5 continued (see legend on pp 16–17)

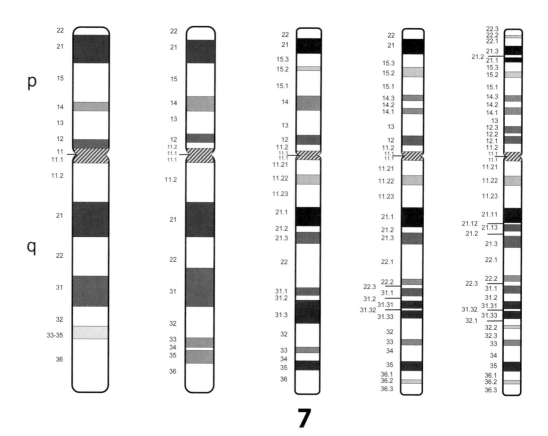

p

q

7

ISCN 2009

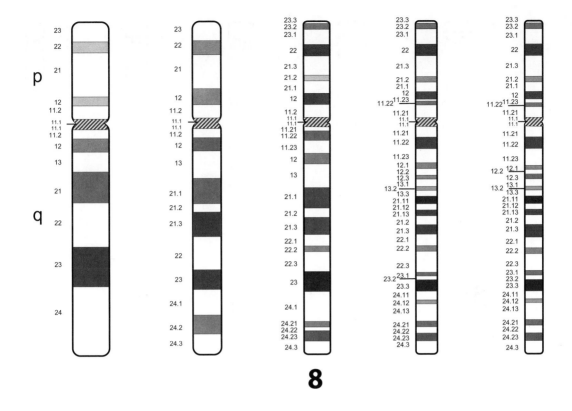

8

Fig. 5 continued (see legend on pp 16–17)

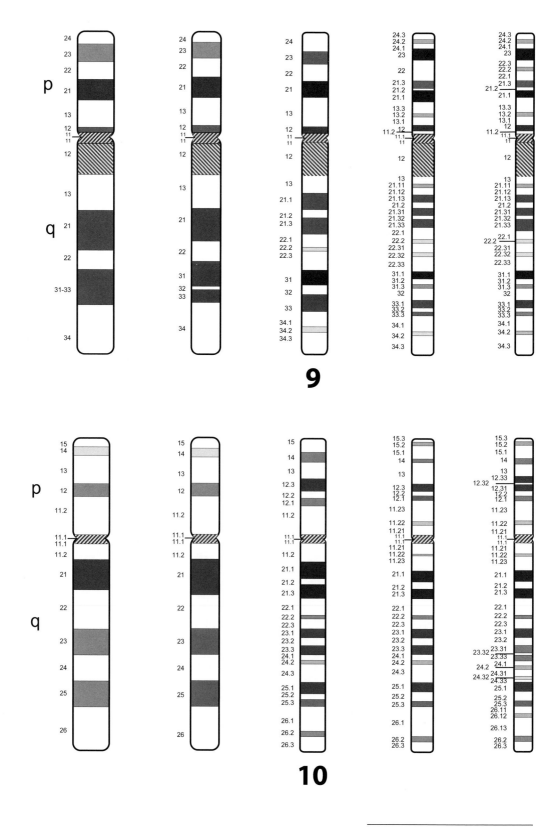

9

10

ISCN 2009

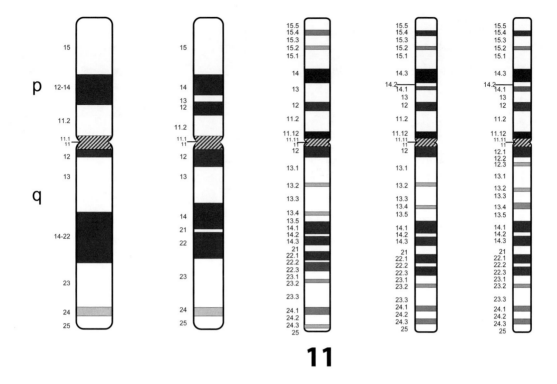

11

Fig. 5 continued (see legend on pp 16–17)

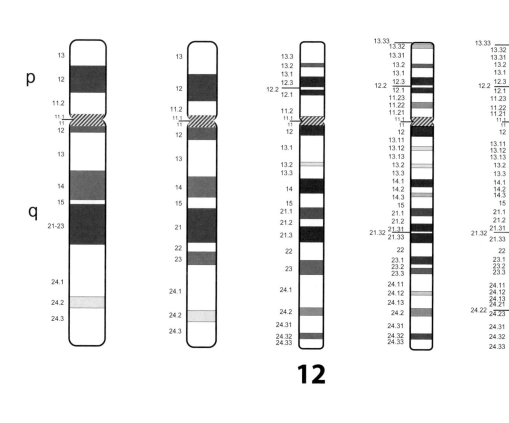

12

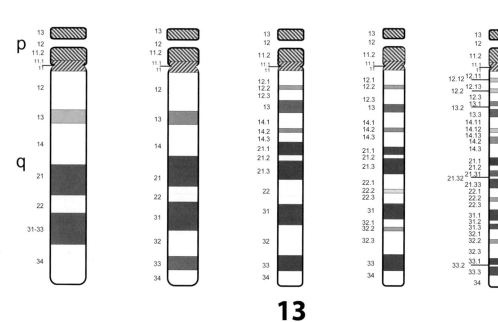

13

ISCN 2009

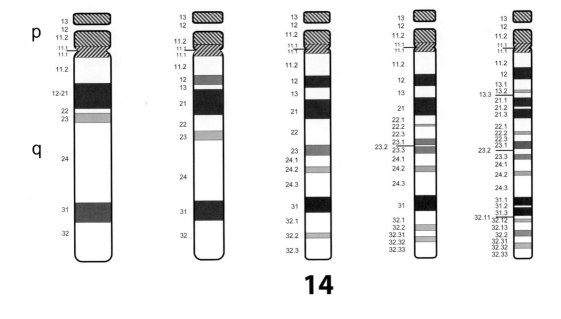

14

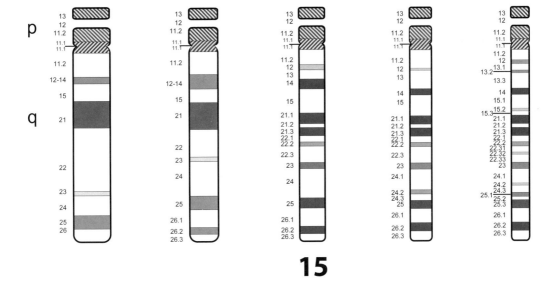

15

Fig. 5 continued (see legend on pp 16–17)

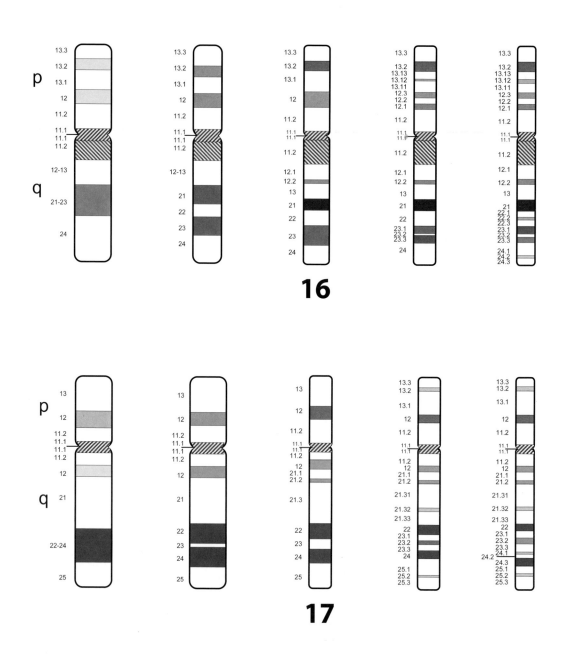

16

17

Fig. 5 continued (see legend on pp 16–17)

ISCN 2009

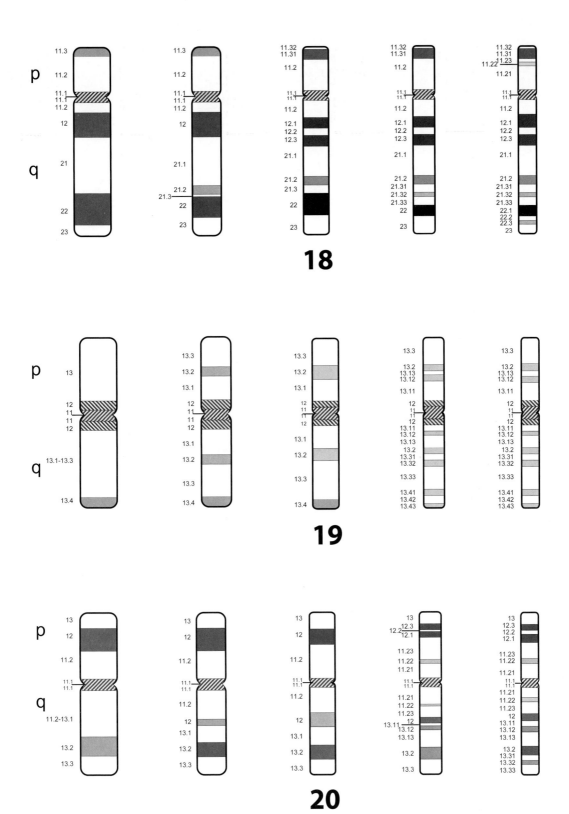

18

19

20

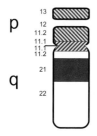
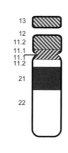
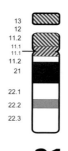
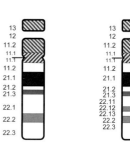

21

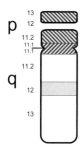
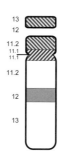
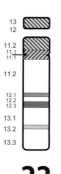
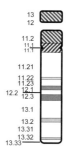

22

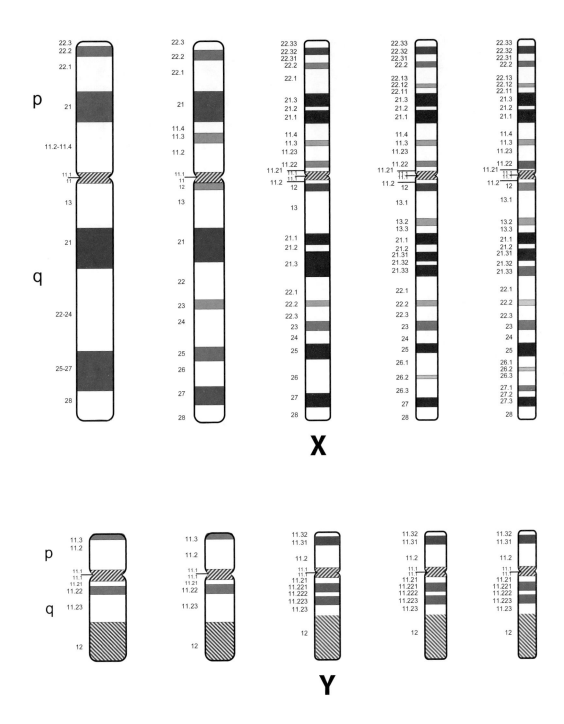

Fig. 5 continued (see legend on pp 16–17)

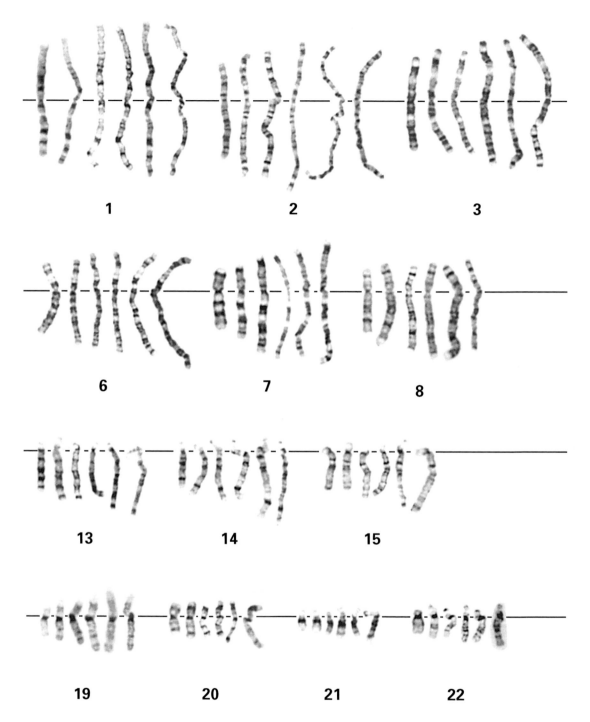

Fig. 6. (a) G-banded chromosomes arranged in increasing order of resolution from approximately the 500- to the 900-band levels. (Courtesy of Dr. E. Magenis).

ISCN 2009

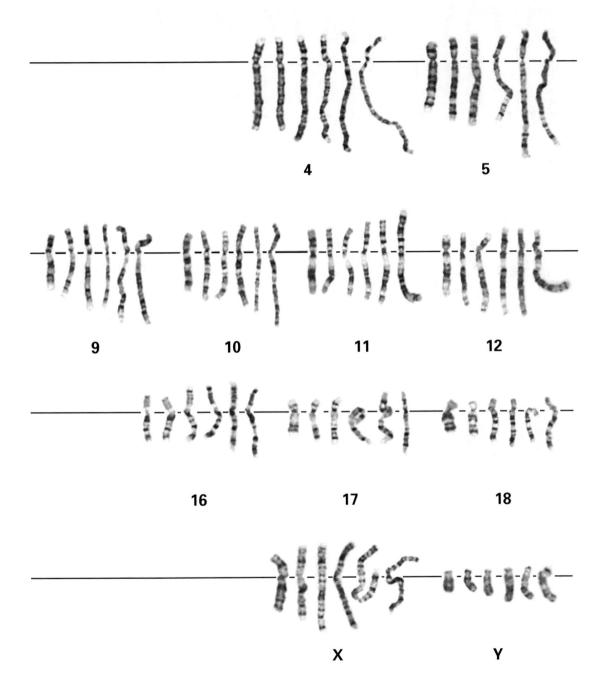

4   5

9   10   11   12

16   17   18

X   Y

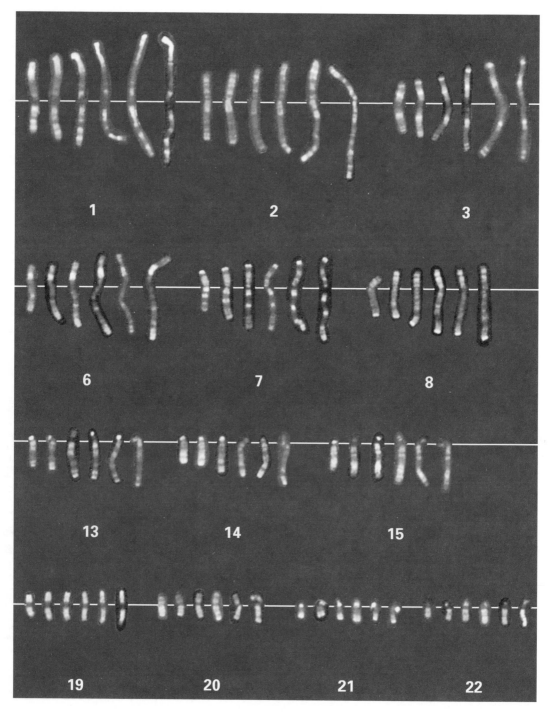

Fig. 6. (b) R-banded chromosomes arranged in increasing order of resolution from approximately the 400- to the 850-band levels. (Courtesy of Dr. E. Magenis.)

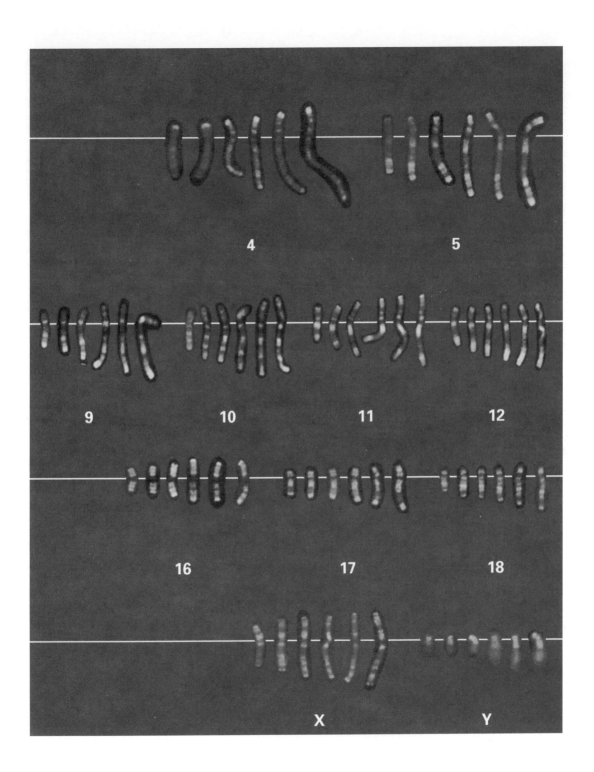

4   5

9   10   11   12

16   17   18

X   Y

3   Symbols and Abbreviated Terms

All symbols and abbreviated terms used in the description of chromosomes and chromosome abnormalities are listed below. Section references are given within parentheses for terms that are defined in greater detail in the text. Abbreviations utilized in describing results obtained by in situ hybridization are given again in Chapter 13 and those utilized in describing microarray results are given again in Chapter 14. When more than one abbreviation is used together, a space is placed between the two abbreviations (e.g. inv dup). When the abbreviation precedes the total number of chromosomes and no parenthesis is present, a space is placed between the abbreviation and the number of chromosomes (e.g. mos 47,XXX[25]/46,XX[5]). There is no space when an abbreviation immediately precedes a parenthesis.

AI	First meiotic anaphase (12.1)
AII	Second meiotic anaphase (12.1)
ace	Acentric fragment (9.2.12, 10.2.1)
add	Additional material of unknown origin (9.2.1)
amp	Denotes an amplified signal (13.2)
approximate sign (~)	Denotes intervals and boundaries of a chromosome segment or number of chromosomes, fragments, or markers (5.2); also used to denote a range of number of copies of a chromosomal region when the exact number cannot be determined (14.2)
arr	Microarray (14.3)
arrow (→ or ->)	From – to, in detailed system (4.3.2.1)
b	Break (10.1.1, 10.2.1)
brackets, angle (< >)	Surround the ploidy level (8.1)
brackets, square ([ ])	Surround number of cells (11.1.2)
c	Constitutional anomaly (4.1, 8.3, 11.3)
cen	Centromere (2.3.2, 4.3.2.1)
cgh	Comparative genomic hybridization (13.7)
chi	Chimera (4.1)
chr	Chromosome (10.2)
cht	Chromatid (10.1)
colon, single (:)	Break, in detailed system (4.3.2.1)
colon, double (::)	Break and reunion, in detailed system (4.3.2.1)
comma (,)	Separates chromosome numbers, sex chromosomes, and chromosome abnormalities (4.1)
con	Connected signals (13.2, 13.4.2)
cp	Composite karyotype (11.1.5)
cx	Complex chromatid interchanges (10.1.1)
decimal point (.)	Denotes sub-bands (2.3.2)
del	Deletion (9.2.2)
der	Derivative chromosome (4.4, 9.2.3, 9.2.17.2, 9.2.17.3)
dia	Diakinesis (12.1)
dic	Dicentric (9.2.4)

dim	Diminished (13.3.1, 13.6)
dip	Diplotene (12.1)
dir	Direct (9.2.5, 9.2.9)
dis	Distal (12.1)
dit	Dictyotene (12.1)
dmin	Double minute (9.2.12)
dn	Designates a chromosome abnormality that has not been inherited (de novo) (4.1)
dup	Duplication (9.2.5)
e	Exchange (10.1.1, 10.2.1)
end	Endoreduplication (4.1)
enh	Enhanced (13.3.1, 13.6)
equal sign (=)	Number of chiasmata (12.1)
fem	Female (12.1)
fib	Extended chromatin/DNA fiber (13.5)
fis	Fission, at the centromere (9.2.6)
fra	Fragile site (7.2, 9.2.7)
g	Gap (10.1.1, 10.2.1)
h	Heterochromatin, constitutive (7.1.1, 7.1.2)
hmz	Homozygous, homozygosity; used when one or two copies of a genome are detected, but previous, known heterozygosity has been reduced to homozygosity through a variety of mechanisms, e.g. loss of heterozygosity (LOH) (14.2)
hsr	Homogeneously staining region (9.2.8)
htz	Heterozygous, heterozygosity (14.2)
i	Isochromosome (9.2.11)
idem	Denotes the stemline karyotype in a subclone (11.1.4)
ider	Isoderivative chromosome (9.2.3)
idic	Isodicentric chromosome (9.2.4, 9.2.11)
inc	Incomplete karyotype (5.4)
ins	Insertion (9.2.9)
inv	Inversion or inverted (9.2.5, 9.2.9, 9.2.10)
ish	In situ hybridization (13.3)
lep	Leptotene (12.1)
MI	First meiotic metaphase (12.1)
MII	Second meiotic metaphase (12.1)
mal	Male (12.1)
mar	Marker chromosome (9.2.12)
mat	Maternal origin (4.1)
med	Medial (12.1)
min	Minute acentric fragment (10.2.1)
minus sign (–)	Loss (4.1, 8.1, 13.2)
ml	Mainline (11.1.3)
mlpa	Multiple ligation-dependent probe amplification (MLPA) (14.1, 14.4)
mn	Modal number (11.2)
mos	Mosaic (4.1)
multiplication sign (×)	Multiple copies of rearranged chromosomes (9.3) or number of copies of a chromosomal region (14.2)
neo	Neocentromere (9.2.13)
nuc	Nuclear (13.4)
oom	Oogonial metaphase (12.1)
or	Alternative interpretation (5.3)
p	Short arm of chromosome (2.3.2)
PI	First meiotic prophase (12.1)
pac	Pachytene (12.1)
parentheses ( )	Surround structurally altered chromosomes and breakpoints (4.1)

A **plus (+)** or **minus (–)** sign is placed before a chromosome or an abnormality designation to indicate additional or missing, normal or abnormal chromosomes, e.g., +21, –7, +der(2); for details, see Section 8.1. The + or – sign placed after a chromosome arm symbol (p or q) may be used in text to indicate an increase or decrease in the length of a chromosome arm (e.g., 4p+, 5q–) but should not be used in the description of karyotypes. See also Sections 9.2.1 and 9.2.2. Variations in length of heterochromatic segments, satellites, and satellite stalks are distinguished from increases or decreases in arm length as a result of other structural alterations by placing a plus or minus sign after the appropriate symbol for these normal variable chromosome features (see Section 7.1). The use of + and – signs in the description of results obtained by in situ hybridization is described in Chapter 13.

When normal chromosomes are replaced by structurally altered chromosomes, the normal ones should not be recorded as missing (see Section 9.1). In the description of karyotypes containing dicentric chromosomes or derivative chromosomes resulting from whole-arm translocations, the abnormal chromosomes by convention replace both normal chromosomes involved in the formation of the dicentric chromosome or derivative chromosome. Thus, in these situations the two missing chromosomes are not specified (see Sections 9.2.4 and 9.2.17.2).

The **multiplication sign (×)** can be used to describe multiple copies of a rearranged chromosome but should not be used to denote multiple copies of normal chromosomes (see Section 9.3).

Uncertainty in chromosome or band designation may be indicated by a **question mark (?)** or an **approximate sign (~)**. The term **or** is used to indicate alternative interpretations of an aberration. For details, see Chapter 5.

The karyotype designations of different clones are separated by a **slant line (/)**. **Square brackets [ ]**, placed after the karyotype description, are used to designate the absolute number of cells in each clone (see Section 11.1.2). In order to distinguish between a *mosaic* (cell lines originating from the same zygote) and a *chimera* (cell lines originating from different zygotes), the triplets **mos** or **chi**, respectively, preceding the karyotype designations, may be used; for example, mos 45,X/46,XX and chi 46,XX/46,XY. In most instances the triplets will be needed only for the initial description in any report; subsequently, the simple karyotype designation may be used. A space should follow mos or chi. All abbreviations that precede a number will have a space that follows. A normal diploid clone, when present, is always listed last, e.g., mos 47,XY,+21/46,XY; mos 47,XXY/46,XY. If there are several abnormal clones, they are presented according to their size; the largest first, then the second largest, and so on, e.g., mos 45,X[15]/47,XXX[10]/46,XX[23]. Likewise, the largest clone in chimeras is presented first, e.g., chi 46,XX[25]/46,XY[10]. When equivalent numbers of cells are found in two cell lines, one of which has a numerical abnormality and the other of which has a structural abnormality, the numerical is listed first, e.g. 45,X[25]/46,X,i(X)(q10)[25]. When both clones have numerical abnormalities they are listed according to the altered autosome number, e.g. 47,XX,+8[25]/47,XX,+21[25]; a clone with a sex chromosome abnormality always comes first, e.g. 47,XXX[25]/47,XX,+21[25]. For order of clone presentation in neoplasia, see Sections 11.1.4 and 11.1.6.

In chimerism secondary to bone marrow transplant, the recipient cell clones are listed first, followed by the donor cell line(s). The recipient and donor cell line(s) are separated by a **double slant line (//)**:

46,XY[3]//46,XX[17]

> Three cells from the male recipient were identified along with 17 cells from the female donor.

46,XY,t(9;22)(q34;q11.2)[4]//46,XX[16]

> Four recipient cells showing a 9;22 translocation were identified along with 16 donor cells.

//46,XX[20]

> All 20 cells were identified as derived from the female donor.

46,XY[20]//

> All 20 cells were identified as derived from the male recipient.

A haploid or polyploid karyotype will be evident from the chromosome number and from the further designations, e.g., 69,XXY. All chromosome changes should be expressed in relation to the appropriate ploidy level (see Sections 8.1 and 9.1), e.g., 70,XXY,+21. An endoreduplicated metaphase cell is indicated by the abbreviation **end** preceding the karyotype designation, e.g., end 46,XX.

When it is known that a particular chromosome involved in an aberration has been inherited from the mother or the father, this may be indicated by the abbreviation **mat** or **pat**, respectively, immediately following the designation of the abnormality, e.g. 46,XX,t(5;6)(q34;q23)mat,inv(14)(q12q31)pat; 46,XX,t(5;6)(q34;q23)mat,inv(14)(q12q31)mat. If it is known that the parents' chromosomes are normal with respect to the abnormality, the abnormality may be designated de novo (**dn**), e.g. 46,XY,t(5;6)(q34;q23)mat,inv(14)(q12q31)dn.

The same rules for designating chromosome aberrations are followed in the description of constitutional and acquired chromosome aberrations. Terms and recommendations related to abnormalities seen in neoplasia are described in Chapter 11. When an acquired chromosome abnormality is found in an individual with a constitutional chromosome anomaly, the latter is indicated by the letter **c** immediately after the constitutional abnormality designation (see Sections 8.3 and 11.3).

In the interest of clarity, complex rearrangements necessitating lengthy descriptions should be written out in full the first time they are used in a report. An abbreviated version may be used subsequently, provided it is clearly defined immediately after the complete notation.

Nomenclature guidelines for meiotic chromosomes are presented in Chapter 12.

4.2 Specification of Breakpoints

The location of any given breakpoint is specified by the band in which that break has occurred. Since it is not possible at present to define band interfaces accurately, a break suspected to be at an interface between two bands is assigned arbitrarily to the higher of the two band numbers, i.e., the number of the band more distal to the centromere.

A given break may sometimes appear to be located in either of two consecutive bands. A similar situation may occur when breaks at or near an interface between two bands are studied with two or more techniques. In this event, the break can be specified by both band numbers separated by the term **or**, e.g., 1q23 or q24, indicating a break in either band 1q23 or band 1q24 (see also Section 5.3). If a break can be localized to a region but not to a particular band, only the region number may be specified, e.g., 1p1. Uncertainty about breakpoint localization may also be indicated by a question mark, e.g., 1p1? (see Section 5.1). If the breakpoint can be assigned only to two adjacent regions, both suspected regions should be indicated, e.g., 1q2 or q3. For the use of the approximate sign to express uncertainty, see Section 5.2.

When an extra copy of a rearranged chromosome is present, the breakpoints are specified only once, at the first time it appears in the karyotype:
48,XX,+1,+der(1)t(1;16)(p13;q13),t(1;16)

4.3 Designating Structural Chromosome Aberrations by Breakpoints and Band Composition

Two systems for designating structural abnormalities exist. One is a **short system** in which the nature of the rearrangement and the breakpoint(s) are identified by the bands or regions in which the breaks occur. Because of the conventions built into this system, the band composition of the abnormal chromosomes can readily be inferred from the information provided in the symbolic description. For very complex abnormalities, especially in tumor cells, the short system may be inadequate or ambiguous, but it will always provide information on all bands involved in the generation of an abnormal chromosome. The other is a **detailed system** which, besides identifying the type of rearrangement, defines each abnormal chromosome in terms of its band composition. The notation used to identify the rearrangement and the method of specifying the breakpoints are common to both systems (see Sections 4.3.1 and 4.3.2).

4.3.1 Short System for Designating Structural Chromosome Aberrations

In this system, structurally altered chromosomes are defined only by their breakpoints. The breakpoints are specified within parentheses immediately following the designation of the type of rearrangement and the chromosome(s) involved. The breakpoints are identified by band designations and are listed in the same order as the chromosomes involved. No semicolon is used between breakpoints in single chromosome rearrangements.

4.3.1.1 Two-Break Rearrangements

When both arms of a single chromosome are involved in a two-break rearrangement, the breakpoint in the short arm is always specified before the breakpoint in the long arm:

46,XX,inv(2)(p21q31)

When two breaks occur within the same arm, the breakpoint more proximal to the centromere is specified first:

46,XX,inv(2)(p13p23)

When two chromosomes are involved, the chromosome having the lowest number is always listed first; however, if one of the rearranged chromosomes is a sex chromosome this is listed first:

46,XY,t(12;16)(q13;p11.1)
46,X,t(X;18)(p11.1;q11.1)

4.3.1.2 Three-Break Rearrangements

An exception to the rule that sex chromosomes and autosomes with the lowest number are specified first involves three-break rearrangements in which part of one chromosome is inserted into another chromosome. In that event, the donor chromosome is listed last, even if it is a sex chromosome or an autosome with a lower number than that of the receptor chromosome:

46,X,ins(5;X)(p14;q21q25)
46,XY,ins(5;2)(p14;q22q32)

When an insertion within a single chromosome occurs, the breakpoint at which the chromosome segment is inserted is always specified first. The remaining breakpoints are specified in the same way as in a two-break rearrangement, i.e., the more proximal breakpoint of the inserted segment is specified first and the more distal one last if the insertion is direct and vice versa if it is inverted:

46,XX,ins(2)(q13p13p23)

> Direct insertion of the short-arm segment between bands 2p13 and 2p23 into the long arm at band 2q13.

46,XX,ins(2)(q13p23p13)

> Inverted insertion of the short-arm segment between bands 2p13 and 2p23 into the long arm at band 2q13. Because the insertion is inverted, band 2p23 is now proximal and band 2p13 distal to the centromere.

For translocations involving three chromosomes, with one breakpoint in each, the rule is still followed that the sex chromosome or autosome with the lowest number is given first. The chromosome listed next is the one that receives a segment from the first chromosome, and the chromosome specified last is the one that donates a segment to the first chromosome listed:

46,XX,t(9;22;17)(q34;q11.2;q22)

> The segment of chromosome 9 distal to 9q34 has been translocated onto chromosome 22 at band 22q11.2, the segment of chromosome 22 distal to 22q11.2 has been translocated onto chromosome 17 at 17q22, and the segment of chromosome 17 distal to 17q22 has been translocated onto chromosome 9 at 9q34.

46,XY,t(X;15;18)(p11.1;p11.1;q11.1)

> The segment of the X chromosome distal to Xp11.1 has been translocated onto chromosome 15 at band 15p11.1, the segment of chromosome 15 distal to 15p11.1 has been translocated onto chromosome 18 at 18q11.1, and the segment of chromosome 18 distal to 18q11.1 has been translocated to Xp11.1.

4.3.1.3 Four-Break and More Complex Rearrangements

Whenever applicable, the guidelines for three-break rearrangements should be used:

46,XX,t(3;9;22;21)(p13;q34;q11.2;q21)

> The segment of chromosome 3 distal to 3p13 has been translocated onto chromosome 9 at 9q34, the segment of chromosome 9 distal to 9q34 has been translocated onto chromosome 22 at 22q11.2, the segment of chromosome 22 distal to 22q11.2 has been translocated onto chromosome 21 at 21q21, and the segment of chromosome 21 distal to 21q21 has been translocated onto chromosome 3 at 3p13.

46,XY,t(5;6)(q13q23;q15q23)

> Reciprocal translocation of two interstitial segments. The segments between bands 5q13 and 5q23 of chromosome 5 and between 6q15 and 6q23 of chromosome 6 have been exchanged.

Unbalanced rearrangements will lead to at least one derivative chromosome and in these situations the use of the symbol **der** to describe the derivative chromosome(s) is recommended (see Section 9.2.3). It will usually not be possible to adequately describe all complex rearrangements with the short system. The detailed system can always be used to describe any abnormality, however complex. Still, it may be necessary to illustrate the rearrangement and/or describe it in words to ensure complete clarity.

4.3.2 Detailed System for Designating Structural Chromosome Aberrations

Structurally altered chromosomes are defined by their band composition. The conventions used in the short system are retained in the detailed system, except that an abbreviated description of the band composition of the rearranged chromosome(s) is specified within the last parentheses, instead of only the breakpoints. It is acceptable to combine the short system (4.3.1) and the detailed system for designating complex karyotypes, especially to describe acquired chromosomal abnormalities.

4.3.2.1 Additional Symbols

A **single colon** (:) is used to indicate a chromosome *break* and a **double colon** (::) to indicate *break and reunion*. In order to avoid an unwieldy description, an **arrow** (→ or ->), meaning *from – to*, is employed. The end of a chromosome arm may be designated either by its band designation or by the symbol **ter** (*terminal*), preceded by the arm designation, i.e., **pter** indicates the end of the short arm and **qter** the end of the long arm. When it is necessary to indicate the *centromere*, the abbreviation **cen** should be used.

4.3.2.2 Designating the Band Composition of a Chromosome

The description starts at the end of the short arm and proceeds to the end of the long arm, with the bands being identified in the order in which they occur in the rearranged chromosome. If the rearrangement is confined to a single chromosome, the chromosome number is not repeated in the band description. If more than one chromosome is involved, however, the bands and chromatid ends are identified with the appropriate chromosome numbers. The aberrations should be listed according to the breakpoints of the derivative chromosome from pter to qter and should not be separated by a comma.

If, owing to a rearrangement, no short-arm segment is present at the end of either arm, the description of the structurally rearranged chromosome starts at the end of the long-arm segment with the lowest chromosome number. However, if a portion of the proximal short arm is present, the description begins with the material on the end of that chromosome arm even if the recipient segment is from a long arm or from a chromosome with a higher or lower chromosome number. For use of the detailed system, see examples in Section 9.2.3.

4.4 Derivative Chromosomes

A *derivative chromosome* is a structurally rearranged chromosome generated by (1) more than one rearrangement within a single chromosome, e.g., an inversion and a deletion of the same chromosome, or deletions in both arms of a single chromosome, or (2) rearrangements involving two or more chromosomes, e.g., the unbalanced product(s) of a translocation. An abnormal chromosome in which no part can be identified is referred to as a marker chromosome (see Section 9.2.12).

Derivative chromosomes are designated **der**. The term always refers to the chromosome(s) that has an intact centromere or neocentromere (Section 9.2.13). The derivative chromosome is specified in parentheses, followed by all aberrations involved in the generation of the derivative chromosome. The aberrations should be listed according to the breakpoints of the derivative chromosome from pter to qter and should not be separated by a comma. For example, der(1)t(1;3)(p32;q21) t(1;11)(q25;q13) specifies a derivative chromosome 1 generated by two translocations, one involving the short arm with a breakpoint in 1p32 and the other involving the long arm with a breakpoint in 1q25.

Various derivative chromosomes and their designations are presented in Section 9.2.3. As an illustration of the way derivative chromosomes can be written, a balanced reciprocal translocation between chromosomes 2 and 5, 46,XX,t(2;5)(q21;q31), has been assumed and is represented by the pachytene diagram in Fig. 7. The derivative chromosomes from such a translocation would be designated der(2) and der(5). Table 4 gives the possible unbalanced gametes resulting from adjacent-1 and adjacent-2 disjunctions and also from four of the 12 possible 3:1 disjunctions, together with the recommended designations of the karyotypes resulting from syngamy between each unbalanced gametic type and a normal gamete. The full karyotype designation needs be written only once in any given publication and then can be abbreviated. A suggested abbreviation for the first designated karyotype in Table 4, for example, would be 46,XX,der(5)mat.

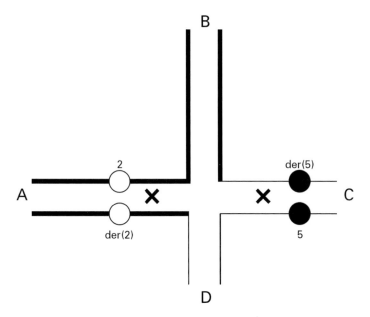

Fig. 7. Pachytene diagram of a t(2;5)(q21;q31) reciprocal translocation heterozygote used to specify the disjunctional possibilities and derivative chromosome combinations given in Table 4. Letters A, B, C, and D designate chromosome ends (telomeres). For the sake of simplicity only those two of the four chromatids that are involved in crossing-over (see Table 4) are indicated. Crosses mark the positions of crossing-over.

4.5   Recombinant Chromosomes

A *recombinant chromosome* is a structurally rearranged chromosome with a new segmental composition resulting from *meiotic* crossing-over between a displaced segment and its normally located counterpart in certain types of structural heterozygotes.

Whereas derivative chromosomes are products of the original rearrangement and segregate at meiosis without further change, recombinant chromosomes arise de novo during gametogenesis in appropriate structural heterozygotes as predictable consequences of crossing-over in a displaced segment.

Recombinant chromosomes are designated by the abbreviation **rec**. The recombinant chromosome is specified in parentheses immediately following the abbreviation. The chromosome designation used is that which indicates the origin of the centromere of the particular recombinant chromosome.

Recombinant chromosomes are most likely to originate from crossing-over in inversion or insertion heterozygotes. To exemplify the method of designating these chromosomes, a maternal pericentric inversion of chromosome 2, 46,XX,inv(2)(p21q31), is shown diagrammatically in Fig. 8. In this case, crossing-over results in a duplication (dup) of 2p in one recombinant chromosome and of 2q in the other. The respective karyotype could be recorded as: 46,XX,rec(2)dup(2p)inv(2)

Table 4. Possible unbalanced gametes derived from segregation of a balanced reciprocal translocation of maternal origin. The pachytene configuration is given in Fig. 7.

Segregation pattern	Schematic segregants	Chromosomal complement of gametes	Karyotype of potential female zygotes
Adjacent-1	AB CB	2, der(5)	46,XX,der(5)t(2;5)(q21;q31)mat
	AD CD	der(2), 5	46,XX,der(2)t(2;5)(q21;q31)mat
Adjacent-2[a]	AB AD	2, der(2)	46,XX,+der(2)t(2;5)(q21;q31)mat,−5
	CD CB	5, der(5)	46,XX,−2,+der(5)t(2;5)(q21;q31)mat
	AB AB	2, 2	46,XX,+2,−5
	AD AD	der(2), der(2)	46,XX,der(2)t(2;5)(q21;q31)mat,+der(2)t(2;5),−5
	CB CB	der(5), der(5)	46,XX,−2,der(5)t(2;5)(q21;q31)mat,+der(5)t(2;5)
	CD CD	5, 5	46,XX,−2,+5
3:1[b]	AB CD CB	2, 5, der(5)	47,XX,+der(5)t(2;5)(q21;q31)mat
	AD	der(2)	45,XX,der(2)t(2;5)(q21;q31)mat,−5
	AD CD CB	der(2), 5, der(5)	47,XX,t(2;5)(q21;q31)mat,+5
	AB	2	45,XX,−5
	AB AD CD	2, der(2), 5	47,XX,+der(2)t(2;5)(q21;q31)mat
	CB	der(5)	45,XX,−2,der(5)t(2;5)(q21;q31)mat
	AB AD CB	2, der(2), der(5)	47,XX,+2,t(2;5)(q21;q31)mat
	CD	5	45,XX,−2

[a] Adjacent-2 disjunction minimally results in the first two unbalanced gametic types shown (AB AD, CD CB). Crossing-over in the interstitial segments between centromeres and points of exchange is necessary for the origin of the remaining four types.

[b] A further eight segregants can occur if there is crossing-over in the interstitial segments, making a total of 12 types of gametes with three chromosomes derived from the translocation quadrivalent.

(p21q31)mat and 46,XX,rec(2)dup(2q)inv(2)(p21q31)mat, specifying, in the first example, a duplication from 2pter to 2p21 and a deletion from 2q31 to 2qter and, in the second example, a duplication from 2q31 to 2qter and a deletion from 2pter to 2p21. Note that, in analogy with the nomenclature for derivative chromosomes, the aberrations following the designation rec are not separated by a comma. The abbreviation rec should only be used when a parental inversion or insertion has been identified. If this is not known, an apparent recombinant chromosome should be written as a derivative. For example, 46,XX,rec(2)dup(2p)inv(2)(p21q31)mat designates a recombinant from a known maternal inversion. 46,XX,der(2)(pter→q31::p21→pter) designates a derivative chromosome with duplication of pter→p21 and deletion of q31→qter. The net imbalance is the same in the two examples, with the first derived from a known inversion carrier.

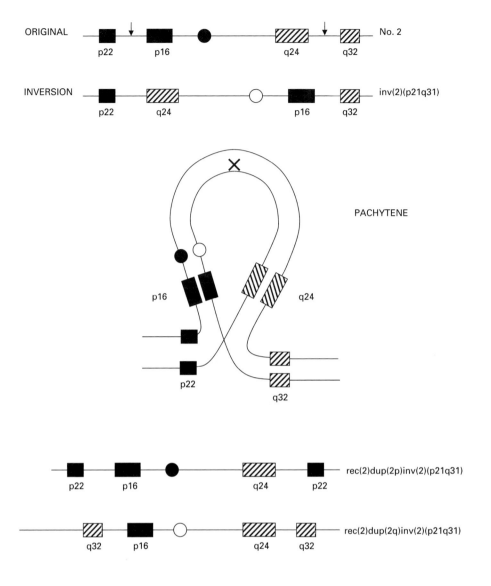

Fig. 8. Diagram of an inv(2)(p21q31)mat pericentric inversion heterozygote. Bands delimiting the breakpoints (arrows on original) are shown as black boxes on the short arm and as hatched boxes on the long arm. In the pachytene diagram, the cross indicates crossing-over within the inversion loop. For the sake of simplicity only those two of the four chromatids that are involved in crossing-over and give rise to the recombinant chromosomes are indicated.

5   Uncertainty in Chromosome or Band Designation

5.1   Questionable Identification

A **question mark (?)** indicates questionable identification of a chromosome or chromosome structure. It is placed either **before the uncertain item**, or it may replace a chromosome, region, or band designation (see also examples in Section 9.2.3).

45,XX,–?21

A missing chromosome, probably No. 21.

47,XX,+?8

An additional chromosome, probably No. 8.

46,XX,del(1)(q2?)

The break in the long arm of chromosome 1 is in region 1q2, but it has not been possible to determine the band within that region.

46,XY,del(1)(q2?3)

The break in the long arm of chromosome 1 is in region 1q2, probably in band 1q23, but this is uncertain.

46,XX,del(1)(q?2)

The break is in the long arm of chromosome 1, probably in region 1q2.

46,XY,del(1)(q?23)

It is uncertain whether the break in the long arm of chromosome 1 is in region 1q2. If so, the break is in band 1q23.

46,XX,del(1)(q?)

The break is in the long arm of chromosome 1, but neither the region nor the band can be identified. This aberration is often described as 1q–, a symbol that may be useful in text but should not be used in karyotype nomenclature.

46,XY,?del(1)(q23)

A possible deletion in chromosome 1, band 1q23, but all items, including the deletion, are uncertain.

46,XX,der(1)?t(1;3)(p22;q13)

The der(1) has probably resulted from a t(1;3). If so, the breaks are in bands 1p22 and 3q13.

5.2   Uncertain Breakpoint Localization or Chromosome Number

An **approximate sign** (~) is used to denote intervals and to express uncertainty about breakpoint localizations in that it indicates the boundaries of a chromosome segment in which the breaks may have occurred.

46,XX,del(1)(q21~24)

> A terminal deletion of the long arm of chromosome 1 with a breakpoint within the segment 1q21–q24, i.e., the breakpoint may be in band 1q21, 1q22, 1q23 or 1q24.

46,XY,dup(1)(q22~24q44)

> A duplication in the long arm of chromosome 1; the proximal breakpoint is in band 1q22, 1q23 or 1q24.

46,XX,t(3;12)(q27~29;q13~15)

> Both breakpoints in this translocation are uncertain; in chromosome 3 the breakpoint may be in bands 3q27, 3q28 or 3q29 and in chromosome 12 in bands 12q13, 12q14 or 12q15.

43~47,XX,…

> The chromosome number is within the interval 43–47.

5.3   Alternative Interpretation

The symbol **or** is used to indicate alternative interpretations of an aberration. Note that there should be a space before and after the symbol.

46,XX,add(19)(p13 or q13)

> Additional material of unknown origin attached to either 19p13 or 19q13.

46,XY,del(8)(q21.1) or i(8)(p10)

> A deletion of the long arm of a chromosome 8 with a breakpoint in 8q21.1 or an isochromosome for the short arm of chromosome 8.

46,XX,t(12;14)(q15;q24) or t(12;14)(q13;q22)

> The two alternative interpretations of the t(12;14) give rise to identical-looking derivative chromosomes. This is in principle a different situation than t(12;14)(q13~15;q22~24), which means that the breakpoint localizations in the t(12;14) are less certain and a variety of combinations are possible.

46,XY,der(1)t(1;10)(q44;q22) or dup(1)(q32q44)

> A rearranged chromosome 1 that may have originated either from a translocation to 1q44 of the distal segment of the long arm of chromosome 10 with a breakpoint in 10q22, or from a duplication of the segment from 1q32 to 1q44.

5.4   Incomplete Karyotype

The symbol **inc** denotes that the karyotype presented is incomplete, usually because of poor chromosome quality. The karyotype thus contains unidentified structural or numerical changes in addition to the abnormalities listed. The symbol **inc** is placed at the end of the nomenclature string, after the description of identifiable abnormalities.

46,XX,del(1)(q21),inc[4]

> It has only been possible to identify a clonally occurring deletion of the long arm of chromosome 1, but there are also additional unidentifiable aberrations. Without the symbol inc the del(1)(q21) would be the sole anomaly in this tumor.

53~57,XY,+1,+3,+6,t(9;22)(q34;q11.2),+21,+3mar,inc[cp10]

> This abnormal karyotype has, in addition to the abnormalities presented that include three marker chromosomes, other changes that could not be identified. cp indicates a composite karyotype from 10 cells (see Section 11.1.5).

Every attempt should be made to present karyotypes in which each abnormality has been identified. The use of inc should therefore be restricted to exceptional situations.

6 Order of Chromosome Abnormalities in the Karyotype

Sex chromosome aberrations are specified first (X chromosome abnormalities are presented before those involving Y), followed by abnormalities of the autosomes listed in numerical order irrespective of aberration type. For each chromosome, numerical abnormalities are listed before structural changes. Multiple structural changes of homologous chromosomes are presented in alphabetical order according to the abbreviated term of the abnormality. For order of clone presentation, see Sections 4.1, 11.1.4 and 11.1.6.

47,X,t(X;13)(q27;q12),inv(10)(p13q22),+21

> The sex chromosome abnormality is presented first, followed by the autosomal abnormalities in chromosome number order, irrespective of whether the aberrations are numerical or structural.

47,Y,t(X;13)(q27;q12),inv(10)(p13q22),+21

> The same karyotype as in the previous example in a male.

46,t(X;18)(p11.1;q11.2),t(Y;1)(q11.2;p13)

> The abnormality involving the X chromosome is listed before that of the Y chromosome.

48,X,t(Y;12)(q11.2;p12),del(6)(q11),+8,t(9; 22)(q34;q11.2),+17,−21,+22

> The translocation involving the Y chromosome is presented first, followed by all autosomal abnormalities in strict chromosome number order.

49,X,inv(X)(p21q26),+3,inv(3)(q21q26.2),+7,+10,−20,del(20)(q11.2),+21

> The inversion of the X chromosome is listed first. The extra chromosome 3 is presented before the inversion of chromosome 3 and the monosomy 20 before the deletion of chromosome 20.

50,XX,+1,+del(1)(p13),+dup(1)(q21q32),+inv(1)(p31q41),+8,r(10)(p12q25),−21

> There are four abnormalities involving different copies of chromosome 1. The numerical change is presented first, followed by the structural aberrations listed in alphabetical order: del, dup, inv.

46,XX,der(8)ins(8;?)(p23;?)del(8)(q22)

> There are two abnormalities involving one chromosome 8. The chromosome 8 is described as a derivative with the structural aberrations listed from the distal p arm to the distal q arm, rather than in alphabetical order, because the insertion and deletion are present on the same derivative chromosome.

Unidentified ring chromosomes (r), marker chromosomes (mar), and double minute chromosomes (dmin) are listed last, in that order:

52,XX,…,+r,+mar,12~20dmin

Derivative chromosomes whose centromere is unknown (see Section 9.2.3) should be placed after all identified abnormalities but before unidentified ring chromosomes, marker chromosomes, and double minute chromosomes:

52,XX,…,+der(?)t(?;6)(?;q16),+r,+mar,5~9dmin

7 Normal Variable Chromosome Features

7.1 Variation in Heterochromatic Segments, Satellite Stalks, and Satellites

Variation refers to the differences in size or staining of chromosomal segments in the population (see Wyandt and Tonk, 2008).

7.1.1 Variation in Length

Variation in length of *heterochromatic segments* (**h**), *stalks* (**stk**) or *satellites* (**s**) should be distinguished from increases or decreases in arm length as a result of other structural alterations by placing a **plus** (+) or **minus** (–) sign after the symbols **h**, **stk** or **s** following the appropriate chromosome and arm designation.

16qh+	Increase in length of the heterochromatin on the long arm of chromosome 16.
Yqh–	Decrease in length of the heterochromatin on the long arm of the Y chromosome.
21ps+	Increase in length of the satellite on the short arm of chromosome 21.
22pstk+	Increase in length of the stalk on the short arm of chromosome 22.
13cenh+pat	Increase in length of the centromeric heterochromatin of the chromosome 13 inherited from the father.
1qh–,13cenh+,22ps+	Decrease in length of the heterochromatin on the long arm of chromosome 1, increase in length of the centromeric heterochromatin on chromosome 13, and large satellites on chromosome 22.
15cenh+mat,15ps+pat	Increase in length of the centromeric heterochromatin on the chromosome 15 inherited from the mother and large satellites on the chromosome 15 inherited from the father.
14cenh+pstk+ps+	Increase in length of the centromeric heterochromatin, the stalk, and the size of satellites on the same chromosome 14.

7.1.2 Variation in Number and Position

The same nomenclature symbols as above are used to describe variation in position of heterochromatic segments, satellite stalks, and satellites:

17ps	Satellites on the short arm of chromosome 17.
Yqs	Satellites on the long arm of the Y chromosome.
9phqh	Heterochromatin in both the short and the long arms of chromosome 9.
9ph	Heterochromatin only in the short arm of chromosome 9.
1q41h	Heterochromatic segment in chromosome 1 at band 1q41.

Duplicated chromosome structures are indicated by repeating the appropriate designation:

| 21pss | Double satellites on the short arm of chromosome 21. |
| 14pstkstk | Double stalks on the short arm of chromosome 14. |

In contrast, the common population inversion variants (see Table 1) are specified by their euchromatic breakpoints.

| inv(9)(p12q13) | Pericentric inversion on chromosome 9. |
| inv(2)(p11.2q13) | Pericentric inversion on chromosome 2. |

7.2   Fragile Sites

Fragile sites (**fra**) associated with a specific disease or phenotype are referred to in Section 9.2.7.

Fragile sites associated with specific chromosome bands can occur as normal variants with no phenotypic consequences. These fragile sites are inherited in a co-dominant Mendelian fashion and may result in chromosome abnormalities such as deletions, multiradial figures, and acentric fragments. While there may be several different types of fragile sites inducible by culturing cells in media containing different components, all these will be covered by a single nomenclature.

fra(10)(q25.2)	A fragile site on chromosome 10 in 10q25.2.
fra(10)(q22.1),fra(10)(q25.2)	Two fragile sites on the same chromosome 10.
fra(10)(q22.1),fra(<u>10</u>)(q25.2)	Two fragile sites on different homologous chromosomes.
fra(10)(q25.2),fra(16)(q22.1)	Two fragile sites on different chromosomes.

8 Numerical Chromosome Abnormalities

8.1 General Principles

A **plus** (+) or **minus** (–) sign is placed before a chromosome to indicate gain or loss of that particular chromosome. The only exception to this rule is the convention to designate constitutional numerical sex chromosome abnormalities by listing all sex chromosomes after the chromosome number, see below.

All numerical changes are expressed in relation to the appropriate ploidy level (see Section 11.2), i.e., in near-haploid cells (chromosome numbers up to 34) in relation to 23, in near-diploid cells (chromosome numbers 35–57) in relation to 46, in near-triploid cells (chromosome numbers 58–80) in relation to 69, in near-tetraploid cells (chromosome numbers 81–103) in relation to 92, and so on.

26,X,+4,+6,+21

> A near-haploid karyotype with two copies of chromosomes 4, 6, and 21, and a single copy of all other chromosomes.

71,XXX,+8,+10

> A near-triploid karyotype with four copies of chromosomes 8 and 10, and three copies of all other chromosomes.

89,XXYY,–1,–3,–5,+8,–21

> A near-tetraploid karyotype with three copies of chromosomes 1, 3, 5, and 21, five copies of chromosome 8, and four copies of all other autosomes.

mos 47,XY,+21[12]/46,XY[18]

> A mosaic karyotype showing two cell lines, one cell line, represented by 12 cells, with trisomy 21 and one normal male cell line, represented by 18 cells. The normal diploid karyotype is written last.

The investigator should select as the reference for the description of the karyotype what is convenient and at the same time biologically meaningful. In such instances, the ploidy level (n, 2n, 3n, etc.) should be given in **angle brackets** <> after the chromosome number.

76~102<4n>,XXXX,…

> The chromosome numbers vary between hypertriploidy and hypertetraploidy. The symbol <4n> indicates that all abnormalities are expressed in relation to the tetraploid level.

58<2n>,XY,+X,+4,+6,+8,+10,+11,+14,+14,+17,+18,+21,+21[10]

> Near-triploid clone with gain of chromosomes listed.

8.2   Sex Chromosome Abnormalities

Constitutional sex chromosome abnormalities are described as follows:

45,X

A karyotype with one X chromosome (Turner syndrome).

47,XXY

A karyotype with two X chromosomes and one Y chromosome (Klinefelter syndrome).

47,XXX

A karyotype with three X chromosomes.

47,XYY

A karyotype with one X chromosome and two Y chromosomes.

48,XXXY

A karyotype with three X chromosomes and one Y chromosome.

mos 47,XXY[10]/46,XY[20]

A mosaic karyotype with one cell line showing two X chromosomes and one Y, found in 10 cells, and a second cell line with a normal diploid male pattern of one X chromosome and one Y chromosome, found in 20 cells.

mos 45,X[25]/47,XXX[12]/46,XX[13]

A mosaic karyotype with two abnormal cell lines, one with monosomy X, found in 25 cells, and one with trisomy X, found in 12 cells. A normal female karyotype was found in 13 cells.

mos 47,XXX[25]/45,X[12]/46,XX[13]

A mosaic karyotype with two abnormal cell lines, one with trisomy X found in 25 cells, and one with monosomy X found in 12 cells. A normal female karyotype was found in 13 cells.

Thus, the constitutional sex chromosome complement is given without the use of plus or minus signs.

Acquired sex chromosome abnormalities are expressed with plus and minus signs as follows:

47,XX,+X

A tumor karyotype in a female with an additional X chromosome.

45,X,–X

A tumor karyotype in a female with loss of one X chromosome.

45,X,–Y

A tumor karyotype in a male with loss of the Y chromosome.

45,Y,–X

A tumor karyotype in a male with loss of the X chromosome.

48,XY,+X,+Y

A tumor karyotype in a male with one additional X and one additional Y chromosome.

Acquired chromosome abnormalities in individuals with a constitutional sex chromosome anomaly can be distinguished with the use of the letter **c** after the constitutional abnormality designation, as illustrated in more detail in Section 11.3.

48,XXYc,+X

Tumor cells with an acquired additional X chromosome in a patient with Klinefelter syndrome.

46,Xc,+X

Tumor cells with an acquired additional X chromosome in a patient with Turner syndrome.

46,XXYc,–X

Tumor cells with an acquired loss of one X chromosome in a patient with Klinefelter syndrome.

44,Xc,–X

Tumor cells with an acquired loss of the X chromosome in a patient with Turner syndrome.

46,Xc,+21

Tumor cells with an acquired extra chromosome 21 in a patient with Turner syndrome.

47,XXX?c

Tumor cells with an uncertain karyotype with an extra X chromosome. The question mark indicates that it is unclear if the extra X is constitutional or acquired.

48,XXY,+mar c

For constitutional markers, there is a space between **mar** and **c**.

8.3   Autosomal Abnormalities

Constitutional and acquired gain or loss of chromosomes are indicated with plus or minus signs.

47,XX,+21

A karyotype with trisomy 21.

48,XX,+13,+21

A karyotype with trisomy 13 and trisomy 21.

45,XX,–22

A karyotype with monosomy 22.

46,XX,+8,−21

A karyotype with trisomy 8 and monosomy 21.

Acquired autosomal abnormalities in individuals with a constitutional anomaly are, as exemplified above for sex chromosome abnormalities, distinguished by the letter **c** (see Section 11.3) after the constitutional abnormality designation.

48,XY,+21c,+21

An acquired extra chromosome 21 in a patient with Down syndrome.

46,XY,+21c,−21

Acquired loss of one chromosome 21 in a patient with Down syndrome.

8.4   Uniparental Disomy

Uniparental disomy, abbreviated **upd**, a condition in which both homologous chromosomes are derived from one parent, may in certain circumstances be identified cytogenetically. See Chapter 14 for UPD example detected by microarray analysis (page 127).

46,XY,upd(15)mat

Male karyotype showing uniparental disomy for a maternally derived chromosome 15.

mos 47,XX,+21[23]/46,XX,upd(21)pat[7]

A mosaic female karyotype consisting of one cell line with uniparental disomy for a paternally derived chromosome 21, identified in 7 cells, and the other with trisomy 21, identified in 23 cells. Note that the trisomic cell line is listed first since it is larger (see Section 4.1).

45,XY,upd der(13;13)(q10;q10)pat

A male karyotype consisting of a single chromosome 13 that is a Robertsonian translocation inherited from the father. Because the father has the same karyotype, this has been interpreted to be uniparental disomy.

9 Structural Chromosome Rearrangements

9.1 General Principles

Structural aberrations, whether constitutional or acquired, should be expressed in relation to the appropriate ploidy level (see Section 11.2), i.e., in near-haploid cells in relation to one chromosome of each type, in near-diploid cells in relation to two chromosomes of each type, in near-triploid cells in relation to three chromosomes of each type, in near-tetraploid cells in relation to four chromosomes of each type, and so on.

69,XXX,del(7)(p11.2)
Two normal chromosomes 7 and one with a deletion of the short arm.

69,XXY,del(7)(q22),inv(7)(p13q22),t(7;14)(p15;q11.1)
No normal chromosome 7: one has a long arm deletion, one has an inversion, and one is involved in a balanced translocation with chromosome 14.

70,XXX,+del(7)(p11.2)
Three normal chromosomes 7 and an additional structurally abnormal chromosome 7 with a deletion of the short arm.

92,XXYY,del(7)(p11.2),t(7;14)(p15;q11.1)
Two normal and two abnormal chromosomes 7: one has a deletion of the short arm, and one is involved in a balanced translocation with chromosome 14.

92,XXYY,del(7)(p11.2),del(7)(q22),del(7)(q34)
One normal chromosome 7 and three with different deletions.

When normal chromosomes are replaced by structurally altered chromosomes, the normal ones should not be recorded as missing.

46,XX,inv(3)(q21q26.2)
An inversion of one chromosome 3. There is no need to indicate that one chromosome 3 is missing, i.e., the karyotype should not be written 46,XX,–3,+inv(3).

45,XX,dic(13;15)(q22;q24)
It is apparent from the symbol dic (see Section 9.2.4) and from the specification of the chromosomes involved that the dicentric chromosome replaces two normal chromosomes. There is thus no need to indicate the missing normal chromosomes.

46,Y,t(X;8)(p22.3;q24.1)

> Male karyotype showing a balanced translocation between the X chromosome and chromosome 8. Note that the normal sex chromosome, in this case a Y, is shown first.

46,XY,der(1)t(1;3)(p22;q13.1)

> The der(1) replaces a normal chromosome 1 and there is no need to indicate the missing normal chromosome. It is obvious from the description that the karyotype contains one normal chromosome 1 and two normal chromosomes 3.

46,XX,ins(1;?)(p22;?)

> Material of unknown origin has been inserted at band p22 in one chromosome 1. The homologous chromosome 1 is normal.

45,XY,−10,der(10)t(10;17)(q22;p12)

> The der(10) replaces a normal chromosome 10; the homologous chromosome 10 is lost. In this situation the missing chromosome 10 must, of course, be indicated.

9.2 Specification of Structural Rearrangements

Examples of structural rearrangements, whether constitutional or acquired, are presented below. Each abnormality is described first with the short system and, when appropriate, also with the detailed system.

9.2.1 Additional Material of Unknown Origin

The symbol **add** (Latin, *additio*) should be used to indicate additional material of unknown origin attached to a chromosome region or band. Such abnormalities have often been described using the symbols t and ?, e.g., t(1;?)(p36;?), but it is only rarely known that the rearranged chromosome has actually resulted from a translocation. The symbol add does not imply any particular mechanism and is therefore recommended.

Additional material attached to a terminal band will always lead to an increase in length of a chromosome arm. Unknown material that replaces a chromosome segment may, depending on the size of the extra material, result in either increase or decrease in the length of the chromosome arm. Designations such as "1p+" or "1p−" may be used in text to describe such abnormal chromosomes, but should not be used in the karyotype.

46,XX,add(19)(p13.3)
46,XX,add(19)(?::p13.3→qter)

> Additional material attached to band 19p13.3, but neither the origin of the extra segment nor the type of rearrangement is known.

46,XY,add(12)(q13)
46,XY,add(12)(pter→q13::?)

> Additional material of unknown origin replaces the segment 12q13qter.

When additional material of unknown origin is attached to both arms of a chromosome and/or replaces more than one segment in a chromosome, the symbol **der** (see Section 9.2.3) should be used.

46,XX,der(5)add(5)(p15.3)add(5)(q23)
46,XX,der(5)(?::p15.3→q23::?)

> Additional material of unknown origin is attached at band 5p15.3 in the short arm and additional material replaces the segment 5q23qter in the long arm.

Unknown material *inserted* in a chromosome arm should be described by the use of the symbols **ins** and **?**.

46,XX,ins(5;?)(q13;?)
46,XX,ins(5;?)(pter→q13::?::q13→qter)

> Material of unknown origin has been inserted into the long arm of chromosome 5 at band 5q13. Use of the symbol add in this situation, i.e., add(5)(q13), would have denoted that unknown material had replaced the segment 5q13qter.

9.2.2 Deletions

The symbol **del** is used to denote both *terminal* and *interstitial* deletions. No arrows are used in the short system to indicate the extent of the deleted segment. This is apparent from the description of the breakpoints. Note that designations such as "5q–" or "del(5q)", which may be useful abbreviations in text, should not be used in karyotypes.

46,XX,del(5)(q13)
46,XX,del(5)(pter→q13:)

> Terminal deletion with a break (:) in band 5q13. The remaining chromosome consists of the entire short arm of chromosome 5 and the part of the long arm lying between the centromere and band 5q13.

46,XX,del(5)(q13q33)
46,XX,del(5)(pter→q13::q33→qter)

> Interstitial deletion with breakage and reunion (::) of bands 5q13 and 5q33. The segment lying between these bands has been deleted.

46,XX,del(5)(q13q13)
46,XX,del(5)(pter→q13::q13→qter)

> Interstitial deletion of a small segment within band 5q13, i.e., both breakpoints are in band 5q13.

46,XY,del(5)(q?)

> Deletion of the long arm of chromosome 5, but it is unclear whether it is a terminal or an interstitial deletion, and also the breakpoints are unknown.

46,Y,del(X)(p21p21)

> Interstitial deletion of a small segment within band Xp21.

Multiple deletions of the same chromosome should be expressed using the symbol **der** (see Section 9.2.3).

9.2.3   Derivative Chromosomes

A *derivative chromosome* (**der**) is a structurally rearranged chromosome generated either by a rearrangement involving two or more chromosomes or by multiple aberrations within a single chromosome. The term always refers to the chromosome that has an intact centromere.

A *recombinant chromosome* (**rec**) is a structurally rearranged chromosome with a new segmental composition resulting from *meiotic* crossing-over and consequently this term should not be used in the description of acquired chromosome abnormalities. If parental karyotypes are unknown or a parental inversion has not been identified, the abnormal chromosome should be designated as a der, not a rec.

46,XX,der(6)(pter→q25.2::p22.2→pter)

When parental karyotypes are known and a parental inversion is identified, rec should be used.

46,XX,rec(6)dup(6p)inv(6)(p22.2q25.2)mat
46,XX,rec(6)(pter→q25.2::p22.2→pter)mat

A **derivative chromosome generated by more than one rearrangement within a chromosome** is specified in parentheses, followed by the type of abnormality. The detailed system is helpful in these cases.

46,XY,der(9)del(9)(p12)del(9)(q31)
46,XY,der(9)(:p12→q31:)

> A derivative chromosome 9 resulting from terminal deletions in both the short and long arms with breakpoints in bands 9p12 and 9q31.

46,XY,der(9)inv(9)(p13p23)del(9)(q22q33)
46,XY,der(9)(pter→p23::p13→p23::p13→q22::q33→qter)

> A derivative chromosome 9 resulting from an inversion in the short arm with breakpoints in 9p13 and 9p23, and an interstitial deletion of the long arm with breakpoints in 9q22 and 9q33.

46,XX,der(7)add(7)(p22)add(7)(q22)
46,XX,der(7)(?::p22→q22::?)

> A derivative chromosome 7 with additional material of unknown origin attached at band 7p22. Similarly, additional material of unknown origin is attached to 7q22, replacing the segment 7q22qter.

46,XX,der(5)add(5)(p15.1)del(5)(q13)
46,XX,der(5)(?::p15.1→q13:)

> A derivative chromosome 5 with additional material of unknown origin attached at 5p15.1 and a terminal deletion of the long arm distal to band 5q13.

A **derivative chromosome resulting from one rearrangement involving two or more chromosomes** is specified in parentheses, followed by the type of abnormality.

46,Y,der(X)t(X;8)(p22.3;q24.1)

> A male showing a derivative X chromosome derived from a translocation between Xp22.3 and 8q24.1.

46,XX,der(1)t(1;3)(p22;q13.1)
46,XX,der(1)(3qter→3q13.1::1p22→1qter)

> The derivative chromosome 1 has resulted from a translocation of the chromosome 3 segment distal to 3q13.1 to the short arm of chromosome 1 at band 1p22. The der(1) replaces a normal chromosome 1 and there is no need to indicate the missing chromosome (see Section 9.1). There are obviously two normal chromosomes 3. The karyotype is unbalanced with loss of the segment 1p22pter and gain of 3q13.1qter.

45,XY,der(1)t(1;3)(p22;q13.1),−3

> The derivative chromosome 1 (same as above) replaces a normal chromosome 1, but there is only one normal chromosome 3. One can presume that it is the der(3) resulting from the t(1;3) that has been lost, but the karyotype cannot make explicit such assumptions.

The term *Philadelphia chromosome* is for historical reasons retained to describe the derivative chromosome 22 generated by the translocation t(9;22)(q34;q11.2). The abbreviation **Ph** (formerly Ph[1]) may be used in text, but not in the description of the karyotype, where der(22)t(9;22)(q34;q11.2) is recommended. Similarly, the derivative chromosome 9 resulting from the t(9;22) is designated der(9)t(9;22) (q34;q11.2).

A **derivative chromosome generated by more than one rearrangement involving two or more chromosomes** is specified in parentheses, followed by all aberrations involved in the generation of the derivative chromosome. The aberrations should be listed according to the breakpoints of the derivative chromosome from pter to qter and should not be separated by a comma.

46,XX,der(1)t(1;3)(p32;q21)t(1;11)(q25;q13)
46,XX,der(1)(3qter→3q21::1p32→1q25::11q13→11qter)

> A derivative chromosome 1 generated by two translocations, one involving the short arm with a breakpoint in 1p32 and the other involving the long arm with a breakpoint in 1q25.

46,XY,der(1)t(1;3)(p32;q21)t(3;7)(q28;q11.2)
46,XY,der(1)(7qter→7q11.2::3q28→3q21::1p32→1qter)

> A derivative chromosome 1 resulting from a translocation of the chromosome 3 segment distal to 3q21 onto 1p32, and a translocation of the segment 7q11.2qter to band 3q28 of the chromosome 3 segment attached to chromosome 1.

46,XY,der(1)t(1;3)(p32;q21)dup(1)(q25q42)
46,XY,der(1)(3qter→3q21::1p32→1q42::1q25→1qter)

> A derivative chromosome 1 resulting from a t(1;3) with a breakpoint in 1p32 and a duplication of the long arm segment 1q25q42.

46,XY,der(9)del(9)(p12)t(9;13)(q34;q11)
46,XY,der(9)(:9p12→9q34::13q11→13qter)

> A derivative chromosome 9 generated by a terminal deletion of the short arm with a breakpoint in 9p12, and by a t(9;13) involving the long arm with a breakpoint in 9q34.

46,XX,der(1)t(1;11)(p32;q13)t(1;3)(q25;q21)
46,XX,der(1)(11qter→11q13::1p32→1q25::3q21→3qter)

> A derivative chromosome 1 generated by two translocations, one involving a breakpoint in 1p32 and 11q13 and the other involving a breakpoint in 1q25 and 3q21. The detailed system describes the derivative 1 from 11qter to 3qter as the aberrations are listed according to the orientation of chromosome 1, from the p arm to the q arm.

47,XY,+mos der(8)r(1;8;17)(p36.3p35;p12q13;q25q25)
47,XY,+mos der(8)r(1;8;17)(::1p36.3→1p35::8p12→8q13::17q25→17q25::)

> Mosaic ring chromosome involving three chromosome segments determined to be a derivative 8 because it retains the 8 centromere. For additional examples of ring chromosomes, see 9.2.15.

46,XX,der(1)del(1)(p22p34)ins(1;17)(p22;q11q25)
46,XX,der(1)(1pter→1p34::17q25→17q11::1p22→1qter)

> A derivative chromosome 1 resulting from an interstitial deletion of the short arm with breakpoints in 1p22 and 1p34, and a replacement of this segment by an insertion of a segment from the long arm of chromosome 17. In such situations, when there are two breakpoints in the recipient chromosome, the proximal one is listed as the point of insertion.

46,XY,der(7)t(2;7)(q21;q22)ins(7;?)(q22;?)
46,XY,der(7)(7pter→7q22::?::2q21→2qter)

> A derivative chromosome 7 in which material of unknown origin has replaced the segment 7q22qter, and the segment 2q21qter from the long arm of chromosome 2 is attached to the unknown chromosome material. By convention, the breakpoint in the derivative chromosome is specified as the point of insertion of the unknown material.

46,XX,der(8)t(8;17)(p23;q21)inv(8)(p22q13)t(8;22)(q22;q12)
46,XX,der(8)(22qter→22q12::8q22→8q13::8p22→8q13::8p22→8p23::17q21→17qter)

> A derivative chromosome 8 resulting from two translocations, one affecting the short arm, one the long arm, with breakpoints at 8p23 and 8q22, respectively, and a pericentric inversion with breakpoints at 8p22 and 8q13.

An *isoderivative chromosome*, abbreviated **ider**, designates an isochromosome formation for one of the arms of a derivative chromosome. The breakpoints are assigned to the centromeric bands p10 and q10 according to the morphology of the isoderivative chromosome (see Section 9.2.11).

46,XX,ider(22)(q10)t(9;22)(q34;q11.2)
46,XX,ider(22)(9qter→9q34::22q11.2→22q10::22q10→22q11.2::9q34→9qter)

> An isochromosome for the long arm of a derivative chromosome 22 generated by a t(9;22), i.e., an isochromosome for the long arm of a Ph chromosome.

46,XY,ider(9)(p10)ins(9;12)(p13;q13q22)
46,XY,ider(9)(9pter→9p13::12q22→12q13::9p13→9p10::9p10→9p13::12q13→
12q22::9p13→9pter)

> An isochromosome for the short arm of a derivative chromosome 9 resulting from an insertion of
> the segment 12q13q22 at band 9p13.

When a derivative chromosome is dicentric and contains one or more additional
abnormalities, the two centromere-containing chromosomes are given within parentheses, separated by a semicolon, followed by the specification of the aberrations.

45,XX,der(5;7)t(5;7)(q22;p13)t(3;7)(q21;q21)
45,XX,der(5;7)(5pter→5q22::7p13→7q21::3q21→3qter)

> A dicentric derivative chromosome. Breakage and reunion have occurred at band 5q22 in the long
> arm of chromosome 5 and at band 7p13 in the short arm of chromosome 7. In addition, the seg-
> ment 3q21qter has been translocated onto the long arm of chromosome 7 at band 7q21.

45,XY,der(5;7)t(3;5)(q21;q22)t(3;7)(q29;p13)
45,XY,der(5;7)(5pter→5q22::3q21→3q29::7p13→7qter)

> A dicentric derivative chromosome composed of chromosomes 5 and 7. The same acentric chro-
> mosome 3 segment as in the previous example is inserted between the long arm of chromosome 5
> and the short arm of chromosome 7.

45,XY,der(5;7)t(3;5)(q21;q22)t(3;7)(q29;p13)del(7)(q32)
45,XY,der(5;7)(5pter→5q22::3q21→3q29::7p13→7q32:)

> The same dicentric derivative chromosome as in the previous example but with an additional ter-
> minal deletion of the long arm of chromosome 7 at band 7q32.

45,XX,der(8;8)(q10;q10)del(8)(q22)t(8;9)(q24.1;q12)
45,XX,der(8;8)(:8q22→8q10::8q10→8q24.1::9q12→9qter)

> A derivative chromosome composed of the long arms of chromosome 8 with material from chro-
> mosome 9 translocated to one arm at band 8q24.1 and a deletion at band 8q22 in the other arm.

When the centromere of the derivative chromosome is not known, but more distal
parts of the chromosome can be recognized, the abnormal chromosome may be designated **der(?)**.

47,XY,+der(?)t(?;9)(?;q22)
47,XY,+der(?)(?→?cen→?::9q22→9qter)

> The distal segment of the long arm of chromosome 9 from band 9q22 has been translocated to a
> centromere-containing derivative chromosome of unknown origin.

47,XX,+der(?)t(?;9)(?;p13)ins(?;7)(?;q11.2q32)
47,XX,+der(?)(9pter→9p13::?→cen→?::7q11.2→7q32::?)

> A derivative chromosome of unknown origin onto which is translocated in its short arm the seg-
> ment of chromosome 9 distal to band 9p13, and which also contains an insertion in the long arm
> of the chromosome 7 segment between bands 7q11.2 and 7q32.

47,XX,+der(?)t(?;9)(?;p13)hsr(?)
47,XX,+der(?)(9pter→9p13::?→cen→?::hsr→?)

> A derivative chromosome of unknown origin with the same translocation in its short arm as in the previous example, and a homogeneously staining region in the long arm.

Derivative chromosomes whose centromeres are unknown should be placed after all identified abnormalities but before unidentified ring chromosomes, marker chromosomes, and double minute chromosomes (see Chapter 6), e.g.:

53,XX,...,+der(?)t(?;9)(?;q22),+r,+mar,dmin

There is usually no need to indicate which homologue is involved in a derivative chromosome because this will be apparent from the karyotype description. If both homologues are involved, this will result in two derivative homologous chromosomes.

46,XX,der(9)del(9)(p12)t(9;22)(q34;q11.2),der(9)t(9;12)(p13;q22)inv(9)(q13q22)

> One der(9) is the result of a deletion of the short arm and a translocation involving the long arm; the other der(9) is the result of a translocation affecting the short arm and a paracentric inversion in the long arm of the homologous chromosome 9. There are two normal chromosomes 12, two normal chromosomes 22, but no normal chromosome 9.

When homologous chromosomes cannot be distinguished within this nomenclature system, one of the numerals may be underlined (single underlining). There is in particular one situation where this may be helpful: When the two homologous chromosomes are involved in identical aberrations resulting in two identical derivative chromosomes.

46,XX,der(1)t(1;3)(p34.3;q21),der(1)t(1;3)(p34.3;q21)

> The two homologous chromosomes 1, as identified by C-band polymorphism, are involved in apparently identical translocations.

46,XX,der(1)t(1;3)(p34.3;q21)[20]/46,XX,der(1)t(1;3)(p34.3;q21)[10]

> The two homologous chromosomes 1 are involved in apparently identical translocations in different cells. Thus, the two abnormalities represent two different clones; the homologous chromosomes 1 in each clone are normal.

Complex rearrangements may give rise to several derivative chromosomes. The breakpoints in the derivative chromosomes generated by the *same rearrangement* need not be repeated in the description of each individual derivative chromosome.

47,XX,t(9;22)(q34;q11.2),+der(22)t(9;22)

> Karyotype with t(9;22) and an additional Ph chromosome. The breakpoints in the extra der(22) need not be repeated.

46,XX,der(1)t(1;3)(p32;q21)inv(1)(p22q21)t(1;11)(q25;q13),der(3)t(1;3),
der(11)t(1;11)

> A balanced complex rearrangement with three derivative chromosomes. The breakpoints of the t(1;3) and the t(1;11), which both contribute to the der(1), are not repeated in the description of der(3) and der(11).

Complex karyotypes involving rearrangements between two or more derivative chromosomes, or where derivative chromosomes are involved in new rearrangements, cannot be described by the short system. The detailed system will be adequate in all such situations. It is acceptable to combine the short system (4.3.1) and the detailed system (4.3.2) for designating complex karyotypes. Whenever doubts remain, the rearrangement should, to avoid ambiguity, be illustrated and described in words.

9.2.4   Dicentric Chromosomes

The symbol **dic** is used to describe *dicentric chromosomes. Isodicentric chromosomes* are designated **idic**. It is apparent from the symbol and from the specification of the chromosome(s) involved that the dicentric chromosome replaces one or two normal chromosomes. There is thus no need to indicate the missing normal chromosome(s) (cf., whole-arm and Robertsonian translocations, Sections 9.2.17.2 and 9.2.17.3). A dicentric chromosome is counted as one chromosome. The term **der** may also be used instead of **dic**, but the combination of **der dic** should not be used.

45,XX,dic(13;13)(q14;q32)
45,XX,dic(13;13)(13pter→13q14::13q32→13pter)

> Breakage and reunion have occurred at bands 13q14 and 13q32 on the two homologous chromosomes 13 to form a dicentric chromosome. There is no normal chromosome 13. If it can be shown that the dicentric chromosome has originated through breakage and reunion of sister chromatids, it may be designated, e.g., dic(13)(q14q32).

45,XX,dic(13;15)(q22;q24)
45,XX,dic(13;15)(13pter→13q22::15q24→15pter)

> A dicentric chromosome with breaks and reunion at bands 13q22 and 15q24. The missing chromosomes 13 and 15 are not indicated since they are replaced by the dicentric chromosome. The karyotype thus contains one normal chromosome 13, one normal chromosome 15, and the dic(13; 15). The resulting net imbalance of this abnormality is loss of the segments distal to 13q22 and 15q24.

46,XX,+13,dic(13;15)(q22;q24)

> A dicentric chromosome with breaks and reunion at bands 13q22 and 15q24 (same as above) has replaced one chromosome 13 and one chromosome 15. There are, however, two normal chromosomes 13, i.e., an additional chromosome 13 in relation to the expected loss due to the dic(13;15). Consequently, the gain is indicated as +13. The karyotype thus contains two normal chromosomes 13, one normal chromosome 15, and the dic(13;15). The resulting net imbalance is partial trisomy for the segment 13pterq22 and loss of the segment 15q24qter.

45,XY,dic(14;21)(p11.2;p11.2)
45,XY,dic(14;21)(14qter→14p11.2::21p11.2→21qter)

> A dicentric chromosome with breaks and reunion at bands 14p11.2 and 21p11.2. The missing chromosomes 14 and 21 are not indicated since they are replaced by the dicentric chromosome. The karyotype thus contains one normal chromosome 14, one normal chromosome 21, and the dic(14;21). The resulting net imbalance of this abnormality is loss of the segments distal to 14p11.2 and 21p11.2. For description of Robertsonian translocations, see Section 9.2.17.3.

47,XY,+dic(17;?)(q22;?)
47,XY,+dic(17;?)(17pter→17q22::?)

> An additional dicentric chromosome composed of one chromosome 17 with a break at band 17q22 and an unknown chromosome with an intact centromere.

46,X,idic(Y)(q12)
46,X,idic(Y)(pter→q12::q12→pter)

> Breakage and reunion have occurred at band Yq12 on sister chromatids to form an isodicentric Y chromosome. The resulting net imbalance is loss of the segment Yq12qter and gain of Ypterq12.

46,XX,idic(21)(q22.3)
46,XX,idic(21)(pter→q22.3::q22.3→pter)

> An isodicentric with breakage and reunion at the terminal ends of two chromosomes 21. There are two copies of the long arm of chromosome 21, joined at q22.3, and one normal chromosome 21, indicated by the 46 count. Even though there are effectively three copies of the chromosome 21 long arm, the normal chromosome 21 is not designated with a (+) sign.

47,XX,+idic(13)(q22)
47,XX,+idic(13)(pter→q22::q22→pter)

> An additional isodicentric chromosome 13. There are two chromosomes 13 and the idic(13). Another example is shown in Section 9.2.11.

47,XY,+idic(15)(q12)
47,XY,+dic(15;15)(q12;q12)
47,XY,+dic(15;15)(pter→q12::q12→pter)

> An additional apparent isodicentric chromosome 15. There are two chromosomes 15 and the idic(15)(q12). This rearrangement has historically been referred to as inv dup(15)(q12). However, because most result from recombination between homologues, dic(15;15)(q12;q12), (or psu dic, see below), would be a more appropriate designation.

Complex dicentric chromosomes will have to be described as derivative chromosomes, see Section 9.2.3.

A *pseudodicentric chromosome* is a dicentric structure in which only one centromere is active. Such chromosomes are abbreviated **psu dic** (similarly, *pseudotricentric*, **psu trc**, etc.), and the segment with the presumptively active centromere, based on the morphology in the majority of cells, is always written first. If the active centromere cannot be determined, the smallest chromosome number is written first.

45,XX,psu dic(15;13)(q12;q12)
45,XX,psu dic(15;13)(15pter→15q12::13q12→13pter)

> A pseudodicentric chromosome has replaced one chromosome 13 and one chromosome 15. The karyotype thus contains one normal chromosome 13, one normal chromosome 15, and the psu dic(15; 13). The centromere of the chromosome mentioned first, i.e., chromosome 15, is the active one.

46,XX,psu idic(20)(q11.2)
46,XX,psu idic(20)(pter→q11.2::q11.2→pter)

> A pseudodicentric chromosome has replaced one chromosome 20, resulting in three copies of 20pterq11.2. The psu idic(20) has one active centromere.

9.2.5 Duplications

The symbol **dup** can be preceded by the triplets **dir** or **inv** to indicate if the duplication is *direct* or *inverted*. However, this is rarely necessary since the orientation of the duplicated segment will be apparent from the order of the bands with respect to the centromere. Note that no arrow is used in the short system to indicate the orientation.

46,XX,dup(1)(q22q25)
46,XX,dup(1)(pter→q25::q22→qter)
> **Direct** duplication of the segment between bands 1q22 and 1q25.

46,XY,dup(1)(q25q22)
46,XY,dup(1)(pter→q25::q25→q22::q25→qter) or (pter→q22::q25→q22::q22→qter)
> **Inverted** duplication of the segment between bands 1q22 and 1q25. Note that only the detailed system will clarify the location of the duplicated segment.

9.2.6 Fission

The symbol **fis** is used to denote centric fission.

47,XY,−10,+fis(10)(p10),+fis(10)(q10)
47,XY,−10,+fis(10)(pter→p10:),+fis(10)(qter→q10:)
> Break in the centromere resulting in two derivative chromosomes composed of the short and long arms, respectively. The breakpoints (:) are assigned to p10 and q10 according to the morphology of the derivative chromosomes.

9.2.7 Fragile Sites

Fragile sites, abbreviated **fra**, may occur as normal variants (see Section 7.2) or be associated with specific diseases and/or phenotypic abnormalities. In both situations the same nomenclature is used.

46,X,fra(X)(q27.3)
> A fragile site in subband Xq27.3 on one X chromosome in a female.

46,Y,fra(X)(q27.3)
> A fragile site in subband Xq27.3 on the X chromosome in a male.

45,fra(X)(q27.3)
> A fragile site in subband Xq27.3 on the X chromosome in a Turner syndrome patient.

47,XY,fra(X)(q27.3)
> A fragile site in subband Xq27.3 on one X chromosome in a Klinefelter syndrome patient.

9.2.8   Homogeneously Staining Regions

The symbol **hsr** is used to describe the presence, but not the size, of a homogeneously staining region in a chromosome arm, segment, or band.

46,XX,hsr(1)(p22)
46,XX,hsr(1)(pter→p22::hsr::p22→qter)

> A homogeneously staining region in band 1p22.

When a chromosome contains multiple hsr or one hsr and another structural change, it is by definition a derivative chromosome and should be designated accordingly (see Section 9.2.3).

46,XX,der(1)hsr(1)(p22)hsr(1)(q31)
46,XX,der(1)(pter→p22::hsr::p22→q31::hsr::q31→qter)

> Two homogeneously staining regions in chromosome 1: one in band 1p22 in the short arm and the other in band 1q31 in the long arm.

46,XY,der(1)del(1)(p21p33)hsr(1)(p21)
46,XY,der(1)(pter→p33::hsr::p21→qter)

> The segment between bands 1p21 and 1p33 is replaced by a homogeneously staining region that may be smaller or larger than the deleted segment. The hsr is by convention assigned to the proximal deletion breakpoint band.

When a homogeneously staining region is located at the interface between segments of different chromosomes involved in a rearrangement, the hsr is assigned to the breakpoints in both chromosomes according to the standard nomenclature for structural chromosome aberrations, i.e., the two chromosomes involved are presented in the first parentheses and the breakpoints in the second.

46,XX,der(1)ins(1;7)(q21;p11.2p21)hsr(1;7)(q21;p11.2)
46,XX,der(1)(1pter→1q21::hsr::7p11.2→7p21::1q21→1qter)

> Insertion of the segment 7p11.2p21 into the long arm of chromosome 1 with breakage and reunion at band 1q21. The derivative chromosome also contains an hsr at the interface between the recipient and donor chromosomes. The hsr is located proximal to the segment inserted from chromosome 7.

46,XX,der(1)ins(1;7)(q21;p11.2p21)hsr(1;7)(q21;p21)
46,XX,der(1)(1pter→1q21::7p11.2→7p21::hsr::1q21→1qter)

> Insertion of the segment 7p11.2p21 into the long arm of chromosome 1 with breakage and reunion at band 1q21. The derivative chromosome also contains an hsr at the interface between the recipient and donor chromosomes. The hsr is located distal to the segment inserted from chromosome 7.

The distinction between homogeneously staining regions and abnormally banded regions, another term that has been used to describe regions containing amplified genes, is ambiguous. So is the distinction between abnormally banded regions and any region of unknown origin within a derivative or marker chromosome. Therefore, until better defined, no symbol to denote abnormally banded regions should be used in karyotype descriptions.

9.2.9 Insertions

The symbol **ins** can be preceded by the triplets **dir** or **inv** to indicate if the original orientation of the inserted segment is retained in its new position or if the orientation is inverted. However, this is rarely necessary since the orientation will be apparent from the order of the bands of the inserted segment with respect to the centromere.

Insertion within a chromosome

46,XX,ins(2)(p13q21q31)
46,XX,ins(2)(pter→p13::q31→q21::p13→q21::q31→qter)

> **Direct insertion**, i.e., dir ins(2)(p13q21q31). The long-arm segment between bands 2q21 and 2q31 has been inserted into the short arm at band 2p13. The original orientation of the inserted segment has been maintained in its new position, i.e., band 2q21 remains more proximal to the centromere than band 2q31.

46,XY,ins(2)(p13q31q21)
46,XY,ins(2)(pter→p13::q21→q31::p13→q21::q31→qter)

> **Inverted insertion**, i.e., inv ins(2)(p13q31q21). The insertion is the same as in the previous example except that the inserted segment has been inverted, i.e., band 2q21 of the inserted segment is now more distal to the centromere than band 2q31. The orientation of the bands within the segment has thus been reversed with respect to the centromere.

Insertion between two chromosomes

46,XX,ins(5;2)(p14;q22q32)
46,XX,ins(5;2)(5pter→5p14::2q32→2q22::5p14→5qter;2pter→2q22::2q32→2qter)

> **Direct insertion**, i.e., dir ins(5;2)(p14;q22q32). The long-arm segment between bands 2q22 and 2q32 has been inserted into the short arm of chromosome 5 at band 5p14. The original orientation of the inserted segment has been maintained in its new position, i.e., 2q22 remains more proximal to the centromere than 2q32. Note that the recipient chromosome is specified first.

46,XY,ins(5;2)(p14;q32q22)
46,XY,ins(5;2)(5pter→5p14::2q22→2q32::5p14→5qter;2pter→2q22::2q32→2qter)

> **Inverted insertion**, i.e., inv ins(5;2)(p14;q32q22). Breakage and reunion have occurred at the same bands as in the previous example, and the insertion is the same except that the inserted segment has been inverted, i.e., band 2q22 is now more distal to the centromere of the recipient chromosome than band 2q32.

46,XX,ins(5;2)(q31;p13p23)
46,XX,ins(5;2)(5pter→5q31::2p13→2p23::5q31→5qter;2pter→2p23::2p13→2qter)

> A direct insertion of bands p13 to p23 from chromosome 2 into band 5q31.

46,XX,ins(5;2)(q31;p23p13)
46,XX,ins(5;2)(5pter→5q31::2p23→2p13::5q31→5qter;2pter→2p23::2p13→2qter)

> An insertion of bands p13 to p23 from chromosome 2 into band 5q31 in an inverted orientation.

9.2.10 Inversions

The symbol **inv** is used. Whether it is a *paracentric* or *pericentric* inversion is apparent from the band designations.

46,XX,inv(3)(q21q26.2)
46,XX,inv(3)(pter→q21::q26.2→q21::q26.2→qter)
> **Paracentric** inversion in which breakage and reunion have occurred at bands 3q21 and 3q26.2.

46,XY,inv(3)(p13q21)
46,XY,inv(3)(pter→p13::q21→p13::q21→qter)
> **Pericentric** inversion in which breakage and reunion have occurred at bands 3p13 and 3q21.

9.2.11 Isochromosomes

The symbol **i** (not iso) is used for isochromosomes and **idic** for isodicentric chromosomes. The breakpoints in isochromosomes are assigned to the centromeric bands p10 and q10 according to the morphology of the isochromosome. See also Section 9.2.4.

46,XX,i(17)(q10)
46,XX,i(17)(qter→q10::q10→qter)
> An isochromosome for the entire long arm of one chromosome 17 and consequently the breakpoint is assigned to 17q10. There is one normal chromosome 17. The shorter designation i(17q) may be used in text but not in the karyotype to describe this isochromosome.

46,X,i(X)(q10)
46,X,i(X)(qter→q10::q10→qter)
> One normal X chromosome and an isochromosome for the long arm of one X chromosome.

47,XY,i(X)(q10)
> A male showing an isochromosome of the long arm of the X chromosome in addition to a normal X and Y.

46,XX,idic(17)(p11.2)
46,XX,idic(17)(qter→p11.2::p11.2→qter)
> An isodicentric chromosome composed of the long arms of chromosome 17 and the short arm materials between the centromeres and the breakpoints in 17p11.2.

46,XX,i(21)(q10)
> An isochromosome of the long arm of chromosome 21 has replaced one chromosome 21. There are two copies of the long arm of chromosome 21 in the isochromosome and one normal copy of chromosome 21. Even though there are effectively three copies of the long arm of chromosome 21, the normal chromosome 21 is not designated with a (+) sign.

Complex isochromosomes, including isoderivative chromosomes, will have to be described as derivative chromosomes, see Section 9.2.3.

9.2.12 Marker Chromosomes

A *marker chromosome* (**mar**) is a structurally abnormal chromosome that cannot be unambiguously identified or characterized by conventional banding cytogenetics. Numerous terms have been used in the literature to describe markers, including "supernumerary marker chromosomes (SMC)," "extra structurally abnormal chromosomes (ESAC)," "supernumerary ring chromosomes (SRC)," and "accessory chromosomes (AC)"; please see Liehr et al. (2004) for a review of these abnormalities. Whenever any part of an abnormal chromosome can be recognized, it is a derivative chromosome (**der**) and can be adequately described by the nomenclature for derivative chromosomes (see Section 9.2.3). For placement of mar in the karyotype, see Chapter 6. In the description of a karyotype, the presence of a mar must be preceded by a **plus sign** (+). No attempt should be made to describe the morphology or size of markers in karyotypes. Thus, the use of, e.g., min mar, A-size mar, acro mar, etc., is discouraged. If such information is relevant, it should be described in words in the text.

47,XX,+mar
> One additional marker chromosome.

47,XX,t(12;16)(q13;p11.2),+mar
> One marker chromosome in addition to t(12;16).

48,X,t(X;18)(p11.2;q11.2),+2mar
> Two marker chromosomes in addition to t(X;18).

47~51,XY,t(11;22)(q24;q12),+1~5mar[cp10]
> In this tumor there are, in addition to a t(11;22), five clonally occurring markers, but not all cells contain all the markers.

When several different markers are clonally present, they may be indicated by an *Arabic* number after the symbol mar, e.g., mar1, mar2, etc. It must be stressed that this does not mean derivation of the marker from chromosome 1, chromosome 2, and so on. Multiple copies of the same marker are indicated by a multiplication sign after the mar designation, e.g., mar1×2 indicates two markers 1; mar1×3 indicates three markers 1, and so on.

48,XX,i(17)(q10),+mar1,+mar2[17]/51,XX,i(17)(q10),+mar1×3,+mar2,+mar3[13]
> There are two different markers (mar1 and mar2) in the clone with 48 chromosomes. The clone with 51 chromosomes has three copies of mar1, one copy of mar2, and in addition a third marker (mar3).

As soon as any part of an abnormal chromosome can be recognized, even if the origin of the centromere is unknown, this abnormal chromosome is referred to as a **der** and not as a mar (see Section 9.2.3).

47,XX,+der(?)t(?;15)(?;q22)
> The centromere of this abnormal chromosome is unknown and hence it is designated der(?), but part of the chromosome is composed of the chromosome 15 segment distal to band 15q22.

Double minutes, abbreviated **dmin**, represent a special kind of acentric structures that should be recorded in the karyotype when present in more than one metaphase cell. Note that the dmin should not be included in the chromosome count, and that the symbol should not be preceded by a plus sign. It is placed after any centric marker. The number of dmin per cell should be presented before the symbol either in absolute numbers or as a mean or a range.

49,XX,…,+3mar,1dmin
>A tumor with one dmin per cell.

49,XY,…,+3mar,~14dmin
>A tumor with approximately 14 dmin per cell.

49,XX,…,+3mar,9~34dmin
>A tumor with 9 to 34 dmin per cell.

Acentric fragments (**ace**) other than dmin, even if present in more than one cell, should not be presented in the karyotype, but must, of course, be recorded in chromosome breakage studies (see Chapter 10).

9.2.13   Neocentromeres

A neocentromere is a functional centromere that has arisen or been activated within a region not known to have a centromere. A chromosome with a neocentromere may be described with the abbreviation **neo** or as a derivative chromosome with the assumption that a new centromere has arisen (or has been activated) within the region(s) from which the chromosome segment was derived.

47,XX,+der(3)(qter→q28:)
>An additional derivative chromosome containing segments 3q28 through 3qter. Because this segment usually does not contain a centromere, this example is a chromosome with a neocentromere. In this example, neo could be used instead of der: 47,XX,+neo(3)(qter→q28:). Also, the location of the neocentromere could be indicated by the symbol neo: 47,XX,+der(3)(qter→q28→neo→q28:). Note that the short system may not be adequate to describe this chromosome.

47,XX,der(3)(:p11→q11:),+neo(3)(pter→p11::q11→q26→neo→q26→qter)
>Chromosome 3 has been replaced by a derivative small chromosome containing the chromosome 3 centromere and by a large chromosome composed of the remaining part of chromosome 3 where a neocentromere has been activated at 3q26.

47,XX,+inv dup(10)(q25qter)
47,XX,+inv dup(10)(qter→q25::q25→qter)
47,XX,+inv dup(10)(qter→q25::q25→q26→neo→q26→qter)
>A supernumerary chromosome that is an inverted duplication of the segments between bands 10q25 and 10qter. Because this segment does not usually contain a centromere, this example is a derivative chromosome with a neocentromere. However, inv dup better describes the nature of the abnormality than neo. The location of the neocentromere could be indicated as shown.

9.2.14 Quadruplications

The symbol **qdp** is used. It is not possible to indicate the orientation(s) of the segment with the short system.

46,XX,qdp(1)(q23q32)
46,XX,qdp(1)(pter→q32::q23→q32::q23→q32::q23→qter)
> Quadruplication of the segment between bands 1q23 and 1q32.

9.2.15 Ring Chromosomes

A ring, designated by the symbol **r**, may be composed of one or several chromosomes.

Ring chromosomes derived from one chromosome

As in other rearrangements affecting a single chromosome, there is no semicolon between the band designations.

46,XX,r(7)(p22q36)
46,XX,r(7)(::p22→q36::)
> Ring chromosome in which breakage and reunion have occurred at bands 7p22 and 7q36. The segments distal to these breakpoints have been deleted.

Ring chromosomes derived from more than one chromosome

Ring chromosomes derived from more than one chromosome may contain one or several centromeres.

Monocentric ring chromosomes are treated as derivative (**der**) chromosomes (see Section 9.2.3). The chromosome that provides the centromere is listed first. The orientation of the acentric segment is apparent from the order of the breakpoints.

46,XX,der(1)r(1;3)(p36.1q23;q21q27)
46,XX,der(1)(::1p36.1→1q23::3q21→3q27::)
> A ring composed of chromosome 1 with breakpoints in 1p36.1 and 1q23, and the acentric segment between bands 3q21 and 3q27 of chromosome 3.

46,XX,der(1)r(1;3)(p36.1q23;q27q21)
46,XX,der(1)(::1p36.1→1q23::3q27→3q21::)
> A ring with the same breakpoints as in the previous example, but the orientation of the acentric segment is reversed.

46,XX,der(1)r(1;?)(p36.1q23;?)
46,XX,der(1)(::1p36.1→1q23::?::)
> A ring composed of chromosome 1 with breakpoints in 1p36.1 and 1q23, and an unknown acentric segment.

If the centromere of the ring chromosome is not known, but segments from other chromosomes contained in the ring are recognized, the ring is designated der(?).

47,XX,+der(?)r(?;3;5)(?;q21q26.2;q13q33)
47,XX,+der(?)(::?→cen→?::3q21→3q26.2::5q13→5q33::)

> In this ring the origin of the centromere is unknown, but the ring contains the acentric segments 3q21 to 3q26.2 and 5q13 to 5q33.

Dicentric or *tricentric ring chromosomes* are designated by the symbol **r** preceded by the triplet **dic** or **trc**.

In *dicentric ring chromosomes* (**dic r**), the sex chromosomes or the autosome with the lowest number is specified first.

47,XX,+dic r(1;3)(p36.1q32;p24q26.2)
47,XX,+dic r(1;3)(::1p36.1→1q32::3p24→3q26.2::)

> A dicentric ring composed of chromosomes 1 and 3 in which 1q32 is fused with 3p24 and 3q26.2 is fused with 1p36.1.

In *tricentric ring chromosomes* (**trc r**), the sex chromosomes or the autosome with the lowest number is specified first. The orientation of the chromosomes will be apparent from the order of the breakpoints.

47,XX,+trc r(1;3;12)(p36.1q32;q26.3p24;p12q23)
47,XX,+trc r(1;3;12)(::1p36.1→1q32::3q26.3→3p24::12p12→12q23::)

> A tricentric ring in which 1q32 is fused with 3q26.3, 3p24 with 12p12, and 12q23 with 1p36.1.

When the origin of the ring is known, the description of the ring is placed in the appropriate chromosome number order:

49,XX,+1,+3,r(7),+8

When the origin of the ring is not known, the presence of the ring, preceded by a plus sign (+), is indicated at the end of the karyotype, but before any other marker chromosome (see Chapter 6).

50,XX,+1,+3,+8,+r
51,XY,+1,+3,+8,+r,+mar

Different rings may be indicated by an *Arabic* number after the symbol r, e.g., r1, r2, etc., whereas several copies of unidentified rings are indicated by the appropriate number before the ring symbol, e.g., 5r.

53,XX,...,+r1,+r2

> Two distinctly different clonally occurring rings. Note that the ring designations r1 and r2 do not mean derivation from chromosomes 1 and 2. When the origin of a ring is known, the appropriate chromosome is placed in parentheses, e.g., r(1), r(2), etc.

53,XY,…,+5r

A total of five rings but it is not known if any of the rings are identical.

9.2.16 Telomeric Associations

The symbol **tas** is used. In telomeric associations between two chromosomes, the sex chromosome or the autosome with the lowest number is specified first. When more than two chromosomes are involved, the 'end' chromosome which has the lowest number, or is one of the sex chromosomes, is specified first, followed by the other chromosomes in the order they are associated with the chromosome listed first. The terminal bands of the chromosomes involved in telomeric association(s) are given in the second parentheses; the orientation of the chromosomes will be apparent from the order in which the bands are listed. Chromosomes involved in telomeric associations are counted as separate chromosomes.

46,XX,tas(12;13)(q24.3;q34)
46,XX,tas(12;13)(12pter→12qter→13qter→13pter)

Association between the telomeric regions of the long arms of chromosomes 12 and 13.

46,XY,tas(X;12;3)(q28;p13q24.3;q29)
46,XY,tas(X;12;3)(Xpter→Xqter→12pter→12qter→3qter→3pter)

Association between the telomeric regions of Xq and 12p, and 12q and 3q.

46,XX,tas(1;X;12;7)(p36.3;q28p22.3;p13q24.3;p22)
46,XX,tas(1;X;12;7)(1qter→1pter→Xqter→Xpter→12pter→12qter→7pter→7qter)

Association between the telomeric regions of 1p and Xq, Xp and 12p, and 12q and 7p.

9.2.17 Translocations

9.2.17.1 Reciprocal Translocations

In *translocations* (**t**) affecting two chromosomes, the sex chromosome or the autosome with the lowest number is always specified first. The same rule is followed in translocations involving three chromosomes, but in these rearrangements the chromosome specified next is the one receiving a segment from the one listed first, and the chromosome specified last is the one donating a segment to the first chromosome listed. Whenever applicable, the same rules should be followed in four-break and more complex balanced translocations. In order to distinguish homologous chromosomes, one of the numerals may be underlined (single underlining).

Two-break rearrangements

46,XY,t(2;5)(q21;q31)
46,XY,t(2;5)(2pter→2q21::5q31→5qter;5pter→5q31::2q21→2qter)

Breakage and reunion have occurred at bands 2q21 and 5q31. The segments distal to these bands have been exchanged.

46,XY,t(2;5)(p12;q31)
46,XY,t(2;5)(5qter→5q31::2p12→2qter;5pter→5q31::2p12→2pter)

> Breakage and reunion have occurred at bands 2p12 and 5q31. The segments distal to these bands have been exchanged.

46,X,t(X;13)(q27;q12)
46,X,t(X;13)(Xpter→Xq27::13q12→13qter;13pter→13q12::Xq27→Xqter)

> Breakage and reunion have occurred at bands Xq27 and 13q12. The segments distal to these bands have been exchanged. Since one of the chromosomes involved in the translocation is a sex chromosome, it is designated first. Note that the correct designation is 46,X,t(X;13) and not 46,XX,t(X;13). Similarly, an identical translocation in a male should be designated 46,Y,t(X;13) and not 46,XY,t(X;13).

46,t(X;Y)(q22;q11.2)
46,t(X;Y)(Xpter→Xq22::Yq11.2→Yqter;Ypter→Yq11.2::Xq22→Xqter)

> A reciprocal translocation between an X chromosome and a Y chromosome with breakpoints at bands Xq22 and Yq11.2.

46,t(X;18)(p11.2;q11.2),t(Y;1)(q11.2;p31)
46,t(X;18)(18qter→18q11.2::Xp11.2→Xqter;18pter→18q11.2::Xp11.2→Xpter),
t(Y;1)(Ypter→Yq11.2::1p31→1pter;Yqter→Yq11.2::1p31→1qter)

> Two reciprocal translocations, each involving one sex chromosome. Breakage and reunion have occurred at bands Xp11.2 and 18q11.2 as well as at bands Yq11.2 and 1p31. Abnormalities of the X chromosome are listed before those of the Y chromosome.

Three-break rearrangements

46,XX,t(2;7;5)(p21;q22;q23)
46,XX,t(2;7;5)(5qter→5q23::2p21→2qter;7pter→7q22::2p21→2pter;5pter→
5q23::7q22→7qter)

> The segment on chromosome 2 distal to 2p21 has been translocated onto chromosome 7 at band 7q22, the segment on chromosome 7 distal to 7q22 has been translocated onto chromosome 5 at 5q23, and the segment of chromosome 5 distal to 5q23 has been translocated onto chromosome 2 at 2p21.

46,X,t(X;22;1)(q24;q11.2;p33)
46,X,t(X;22;1)(Xpter→Xq24::1p33→1pter;22pter→22q11.2::Xq24→Xqter;
22qter→22q11.2::1p33→1qter)

> The segment on one X chromosome distal to Xq24 has been translocated onto chromosome 22 at band 22q11.2, the segment distal to 22q11.2 has been translocated onto chromosome 1 at band 1p33, and the segment distal to 1p33 has been translocated onto the X chromosome at band Xq24.

46,XX,t(2;7;7)(q21;q22;p13)
46,XX,t(2;7;7)(2pter→2q21::7p13→7pter;7pter→7q22::2q21→2qter;7qter→7q22::
7p13→7qter)

> The segment on chromosome 2 distal to 2q21 has been translocated onto chromosome 7 at band 7q22, the segment on chromosome 7 distal to 7q22 has been translocated onto the homologous chromosome 7 at band 7p13, and the segment distal to 7p13 on the latter chromosome has been

translocated onto chromosome 2 at 2q21. Underlining is used only to emphasize that the chromosomes are homologous. However, this is usually not necessary since if the same chromosome 7 had been involved, the resulting chromosome 7 would have to be described as a derivative chromosome.

Four-break and more complex rearrangements

46,XX,t(3;9;22;21)(p13;q34;q11.2;q21)
46,XX,t(3;9;22;21)(21qter→21q21::3p13→3qter;9pter→9q34::3p13→3pter;
22pter→22q11.2::9q34→9qter;21pter→21q21::22q11.2→22qter)

> The segment of chromosome 3 distal to 3p13 has been translocated onto chromosome 9 at 9q34; the segment of chromosome 9 distal to 9q34 has been translocated onto chromosome 22 at 22q11.2; the segment of chromosome 22 distal to 22q11.2 has been translocated onto chromosome 21 at 21q21; and the segment of chromosome 21 distal to 21q21 has been translocated onto chromosome 3 at 3p13.

46,XX,t(3;9;9;22)(p13;q22;q34;q11.2)
46,XX,t(3;9;9;22)(22qter→22q11.2::3p13→3qter;9pter→9q22::3p13→3pter;
9pter→9q34::9q22→9qter;22pter→22q11.2::9q34→9qter)

> Four-break rearrangement involving the two homologous chromosomes 9. The segment on chromosome 3 distal to 3p13 has been translocated onto chromosome 9 at band 9q22, the segment on chromosome 9 distal to 9q22 has been translocated onto the homologous chromosome 9 at 9q34, the segment on the latter chromosome 9 distal to 9q34 has been translocated onto chromosome 22 at 22q11.2, and the segment on chromosome 22 distal to 22q11.2 has been translocated onto chromosome 3 at 3p13.

46,XY,t(5;6)(q13q23;q15q23)
46,XY,t(5;6)(5pter→5q13::6q15→6q23::5q23→5qter;6pter→6q15::5q13→
5q23::6q23→6qter)

> Four-break rearrangement involving two chromosomes. The segment between bands 5q13 and 5q23 in chromosome 5 and the segment between bands 6q15 and 6q23 in chromosome 6 have been exchanged.

46,XX,t(5;14;9)(q13q23;q24q21;p12p23)
46,XX,t(5;14;9)(5pter→5q13::9p12→9p23::5q23→5qter;14pter→14q21::5q13→
5q23::14q24→14qter;9pter→9p23::14q21→14q24::9p12→9qter)

> Reciprocal six-break translocation of three interstitial segments. The segment between bands 5q13 and 5q23 on chromosome 5 has replaced the segment between bands 14q21 and 14q24 on chromosome 14; the segment 14q21q24 has replaced the segment between bands 9p12 and 9p23 on chromosome 9; and the segment 9p12p23 has replaced the segment 5q13q23. The orientations of the segments in relation to the centromere are apparent from the order of the bands. Thus, the segment 14q21q24 is inverted.

The *derivative chromosomes* produced by reciprocal translocations should be described using the conventions outlined in Section 9.2.3.

9.2.17.2 Whole-Arm Translocations

Whole-arm translocations can be adequately described by assigning the breakpoints to the centromeric bands p10 and q10 according to the morphology of the abnormal chromosomes. In *balanced whole-arm exchanges*, the breakpoint in the chromosome which has the lowest number, or is a sex chromosome, is assigned to p10.

46,XY,t(1;3)(p10;q10)
46,XY,t(1;3)(1pter→1p10::3q10→3qter;3pter→3p10::1q10→1qter)

> Reciprocal whole-arm translocation in which the short arm of chromosome 1 has been fused at the centromere with the long arm of chromosome 3 and the long arm of chromosome 1 has been fused with the short arm of chromosome 3.

46,XY,t(1;3)(p10;p10)
46,XY,t(1;3)(1pter→1p10::3p10→3pter;1qter→1q10::3q10→3qter)

> Reciprocal whole-arm translocation in which the short arms of chromosomes 1 and 3 and the long arms of these chromosomes, respectively, have been fused at the centromeres.

In the description of karyotypes containing derivative chromosomes resulting from *unbalanced whole-arm translocations* (see Section 9.2.3), the derivative chromosome (**der**) by convention replaces the two normal chromosomes involved in the translocation. Thus, the two missing normal chromosomes are not specified. The imbalance(s) will be obvious from the karyotype designation.

45,XX,der(1;3)(p10;q10)
45,XX,der(1;3)(1pter→1p10::3q10→3qter)

> A derivative chromosome consisting of the short arm of chromosome 1 and the long arm of chromosome 3. The missing chromosomes 1 and 3 are not indicated since they are replaced by the derivative chromosome. The karyotype thus contains one normal chromosome 1, one normal chromosome 3, and the der(1;3). The resulting net imbalance of this abnormality is monosomy for the long arm of chromosome 1 and monosomy for the short arm of chromosome 3.

46,XX,+1,der(1;3)(p10;q10)

> A derivative chromosome consisting of the short arm of chromosome 1 and the long arm of chromosome 3 (same as above) has replaced one chromosome 1 and one chromosome 3. There are, however, two normal chromosomes 1, i.e., an additional chromosome 1 in relation to the expected loss due to the der(1;3). Consequently, this gain is indicated as +1. The karyotype thus contains two normal chromosomes 1, one normal chromosome 3, and the der(1;3). The resulting net imbalance is trisomy for 1p and monosomy for 3p.

46,XX,der(1;3)(p10;q10),+3

> A derivative chromosome consisting of the short arm of chromosome 1 and the long arm of chromosome 3 (same as above) has replaced one chromosome 1 and one chromosome 3. There are, however, two normal chromosomes 3, i.e., an additional chromosome 3 in relation to the expected loss due to the der(1;3). Consequently, this gain is indicated as +3. The karyotype thus contains one normal chromosome 1, two normal chromosomes 3, and the der(1;3). The resulting net imbalance is monosomy for 1q and trisomy for 3q.

47,XX,+der(1;3)(p10;q10)

> An extra derivative chromosome consisting of the short arm of chromosome 1 and the long arm of chromosome 3 (same as above). There are thus two normal chromosomes 1, two normal chromosomes 3, and the der(1;3). The resulting net imbalance is trisomy for 1p and trisomy for 3q.

44,XY,−1,der(1;3)(p10;q10)

> A derivative chromosome consisting of the short arm of chromosome 1 and the long arm of chromosome 3 (same as above) has replaced one chromosome 1 and one chromosome 3. There is, however, no normal chromosome 1, indicated as −1 in relation to the expected presence of one chromosome 1 due to the der(1;3). The karyotype thus contains no chromosome 1, one normal chromosome 3, and the der(1;3). The resulting net imbalance is nullisomy for 1q, monosomy for 1p, and monosomy for 3p.

9.2.17.3 Robertsonian Translocations

These special types of translocations originate through translocation of the acrocentric chromosomes 13–15 and 21–22. The breakpoints mostly occur in the short arms, resulting in dicentric chromosomes. Breaks may also occur in one short arm and one long arm of the participating chromosomes, resulting in monocentric rearrangements. Usually there is simultaneous loss of the remaining short arms. Either **rob** or **der** can adequately describe these whole-arm translocations.

45,XX,der(13;21)(q10;q10)
45,XX,rob(13;21)(q10;q10)

> Breakage and reunion have occurred at band 13q10 and band 21q10 in the centromeres of chromosomes 13 and 21. The derivative chromosome has replaced one chromosome 13 and one chromosome 21 and there is no need to indicate the missing chromosomes. The karyotype thus contains one normal chromosome 13, one normal chromosome 21, and the der(13;21). The resulting net imbalance is loss of the short arms of chromosomes 13 and 21.

46,XX,der(13;21)(q10;q10),+21
46,XX,rob(13;21)(q10;q10),+21

> A derivative chromosome consisting of the long arm of chromosome 13 and the long arm of chromosome 21 (same as above) has replaced one chromosome 13 and one chromosome 21. There are, however, two normal chromosomes 21, i.e., an additional chromosome 21 in relation to the expected loss due to the der(13;21). Consequently, this gain is indicated as +21. The karyotype thus contains one normal chromosome 13, two normal chromosomes 21, and the der(13;21). The resulting net imbalance is loss of the short arm of chromosome 13 and trisomy for the long arm of chromosome 21.

46,XX,+13,der(13;21)(q10;q10)
46,XX,+13,rob(13;21)(q10;q10)

> A derivative chromosome consisting of the long arm of chromosome 13 and the long arm of chromosome 21 (same as above) has replaced one chromosome 13 and one chromosome 21. There are, however, two normal chromosomes 13, i.e., an additional chromosome 13 in relation to the expected loss due to the der(13;21). Consequently, this gain is indicated as +13. The karyotype thus contains two normal chromosomes 13, one normal chromosome 21, and the der(13;21). The resulting net imbalance is loss of the short arm of chromosome 21 and trisomy for the long arm of chromosome 13.

If only a single chromosome is involved in the rearrangement, the extra chromosome is indicated by the 46 count in the presence of a whole-arm rearrangement and the addition of a normal chromosome.

46,XX,+21,der(21;21)(q10;q10)
46,XX,+21,rob(21;21)(q10;q10)

> A derivative chromosome composed of the long arms of chromosome 21. There are two copies of the long arm of chromosome 21 in the derivative chromosome and one normal chromosome 21, indicated by the 46 count. The normal chromosome 21 is designated with a (+) sign.

The abbreviation rob should not be used in the description of acquired abnormalities.

If it is proven that the derivative chromosome resulting from a whole-arm translocation is *dicentric*, i.e., the breakpoints have been assigned to p11.2 or q11.2, the abbreviation **dic** should be used and the dicentric chromosome should be described accordingly (see Section 9.2.4).

9.2.17.4 Jumping Translocations

These can be adequately described by the standard nomenclature for translocations. The clones are presented in the same order as unrelated clones, i.e., in order of decreasing frequency (see Section 11.1.6).

46,XX,t(4;7)(q35;q11.2)[6]/46,XX,t(1;7)(p36.3;q11.2)[4]/46,XX,t(7;9)(q11.2; p24)[3]

> Three clonal translocations involving band 7q11.2. The segment 7q11.2qter is translocated to bands 1p36.3, 4q35, and 9p24.

9.2.18   Tricentric Chromosomes

The symbol **trc** is used. The 'end' chromosome which has the lowest number, or is one of the sex chromosomes, is specified first. The other chromosomes are listed in the order they are attached to the chromosome listed first. The orientation of the chromosomes will be apparent from the order of the breakpoints specified in the second parentheses. A tricentric chromosome is counted as one chromosome.

44,XX,trc(4;12;9)(q31.2;q22p13;q34)
44,XX,trc(4;12;9)(4pter→4q31.2::12q22→12p13::9q34→9pter)

> A tricentric chromosome in which band 4q31.2 is fused with 12q22 and 12p13 is fused with 9q34.

9.2.19   Triplications

The symbol **trp** is used. It is not possible to indicate the orientation(s) of the segment in the short system, but this can be done with the detailed system.

46,XX,trp(1)(q21q32)
46,XX,trp(1)(pter→q32::q21→q32::q21→qter)

> Direct triplication of the segment between bands 1q21 and 1q32, one of several possible orientations of the triplications of this segment.

46,XX,inv trp(1)(q32q21)
46,XX,inv trp(1)(pter→q32::q32→q21::q21→qter)

> Inverted triplication of the segment between bands 1q21 and 1q32.

9.3   Multiple Copies of Rearranged Chromosomes

The **multiplication sign** (×) can be used to describe two or more copies of a structurally rearranged chromosome. The number of copies (×2, ×3, etc.) should be placed after the abnormality. The multiplication sign should not be used to denote multiple copies of normal chromosomes.

46,XX,del(6)(q13q23)×2

> Two deleted chromosomes 6 with breakpoints at bands 6q13 and 6q23, and no normal chromosome 6. Since the two abnormal chromosomes replace the two normal chromosomes, there is no need to indicate the missing normal chromosomes.

48,XY,+del(6)(q13q23)×2

> Two normal chromosomes 6 plus two additional deleted chromosomes 6 with breakpoints at bands 6q13 and 6q23.

47,XX,del(6)(q13q23)×2,+del(6)(q13q23)

> There are three copies of a deleted chromosome 6 and no normal chromosome 6, i.e., two of the deleted chromosomes replace the two normal chromosomes 6. Note that the supernumerary deleted chromosome 6 has to be preceded by a plus sign.

48,XX,del(6)(q13q23)×2,+7,+7

> Two deleted chromosomes 6 replace the two normal chromosomes 6; in addition, there are two extra chromosomes 7.

48,XX,t(8;14)(q24.1;q32),+der(14)t(8;14)×2

> A balanced t(8;14) plus two additional copies of the derivative chromosome 14.

92,XXXX,t(8;14)(q24.1;q32)×2

> A tetraploid clone with two balanced t(8;14). The two derivative chromosomes 8 and 14 replace two normal chromosomes 8 and 14. Thus, there are two normal chromosomes 8 and 14.

94,XXYY,t(8;14)(q24.1;q32)×2,+der(14)t(8;14)×2

> A hypertetraploid clone with two balanced t(8;14) plus two additional copies of the derivative chromosome 14. There are two normal chromosomes 8 and 14.

93,XXXX,t(8;14)(q24.1;q32)×2,der(14)t(8;14)×2,+der(14)t(8;14)

> A hypertetraploid clone with two balanced t(8;14) and three extra copies of the derivative chromosome 14, i.e., there are in total five der(14), four of which replace the normal chromosomes 14; consequently there is no normal chromosome 14.

94,XXYY,t(8;14)(q24.1;q32)×2,+14,der(14)t(8;14)×2,+der(14)t(8;14)

A hypertetraploid clone with two balanced t(8;14), three extra copies of the derivative chromosome 14, and one normal chromosome 14.

47,XX,+8,i(8)(q10)×2
47,XX,i(8)(q10),+i(8)(q10)

Alternative descriptions of the same chromosome complement with one normal chromosome 8 and two copies of an isochromosome for the long arm of chromosome 8.

10   Chromosome Breakage

This section provides a nomenclature for the chromatid and chromosome aberrations that may be observed in, for example, constitutional chromosome breakage syndromes or following clastogenic exposure. Since many aberrations of this kind are scored on unbanded material, recommendations are given first for non-banded preparations; these are followed by recommendations for banded preparations.

10.1   Chromatid Aberrations

10.1.1   Non-Banded Preparations

A *chromatid* (**cht**) aberration involves only one chromatid in a chromosome at a given locus.

A *chromatid gap* (**chtg**) is a non-staining region (achromatic lesion) of a single chromatid in which there is minimal misalignment of the chromatid.

A *chromatid break* (**chtb**) is a discontinuity of a single chromatid in which there is a clear misalignment of one of the chromatids.

A *chromatid exchange* (**chte**) is the result of two or more chromatid lesions and the subsequent rearrangement of chromatid material. Exchanges may be between chromatids of different chromosomes (*interchanges*) or between or within chromatids of one chromosome (*intrachanges*). In the case of interchanges, it will generally be sufficient to indicate whether the configuration is *triradial* (**tr**) when there are three arms to the pattern, *quadriradial* (**qr**) when there are four, or *complex* (**cx**) when there are more than four. The number of centromeres might be indicated within parentheses (1 cen, 2 cen, etc.). When necessary, exchanges may be classified in more detail. *Asymmetrical* exchanges inevitably result in the formation of an acentric fragment, whereas *symmetrical* ones do not. In *complete exchanges* all the broken ends are rejoined, but not in *incomplete* ones. In asymmetrical exchanges, the incompleteness may be proximal when the broken ends nearest the centromere are not rejoined or distal when the ends farthest from the centromere are not rejoined. Intra-arm events include duplications, deletions, paracentric inversions, and isochromatid breaks showing sister reunion. It should be noted that these terms are only descriptive and do not imply knowledge of the origin of the aberrations.

Sister chromatid exchange, detectable only by special staining methods, results from the interchange of homologous segments between two chromatids of one chromosome. The abbreviation **sce** can be used to describe this event.

10.1.2   Banded Preparations

Some chromatid aberrations can be defined more precisely or can be recognized with certainty only in banded preparations; e.g., a *chromatid deletion* (**cht del**) is the absence of a banded sequence from only one of the two chromatids of a single chromosome. A *chromatid inversion* (**cht inv**) is the reversal of a banded sequence of only one of the two chromatids of a single chromosome. Both are subclasses of *chromatid exchanges* (**chte**).

Where it is desired to specify the location of a chromatid aberration, the appropriate symbol can be followed by the band designation, e.g.:

chtg(4)(q25)	Chromatid gap in chromosome 4 at band 4q25.
chtb(4)(q25)	Chromatid break in chromosome 4 at band 4q25.
chte(4;10)(q25;q22)	Chromatid exchange involving chromosomes 4 and 10 at bands 4q25 and 10q22.
cht del(1)(q12q25)	Chromatid deletion in chromosome 1 with loss of the segment between bands 1q12 and 1q25.
cht inv(1)(q12q25)	Chromatid inversion in chromosome 1 with reversal of the segment between bands 1q12 and 1q25.
sce(4)(q25q33)	Sister chromatid exchanges in chromosome 4 at bands 4q25 and 4q33.

10.2   Chromosome Aberrations

10.2.1   Non-Banded Preparations

A *chromosome* (**chr**) aberration involves both chromatids of a single chromosome at the same locus.

A *chromosome gap* (**chrg**) is a non-staining region (achromatic lesion) at the same locus in both chromatids of a single chromosome in which there is minimal misalignment of the chromatids. The term *chromosome gap* is synonymous with *isolocus gap* and *isochromatid gap*.

A *chromosome break* (**chrb**) is a discontinuity at the same locus in both chromatids of a single chromosome, giving rise to an acentric fragment and an abnormal monocentric chromosome. This fragment is therefore a particular type of acentric fragment (**ace**), and chrb should be used only when the morphology indicates that the fragment is the result of a single event. The term *chromosome break* is synonymous with *isolocus break* and *isochromatid break*.

A *chromosome exchange* (**chre**) is the result of two or more chromosome lesions and the subsequent relocation of both chromatids of a single chromosome to a new position on the same or on another chromosome. It may be symmetrical (e.g., reciprocal translocation) or asymmetrical (e.g., dicentric formation).

A *minute* (**min**) is an acentric fragment smaller than the width of a single chromatid. It may be single or double. In the special situation when *double minutes* are present clonally in tumor cells, the abbreviation **dmin** is used; see Section 9.2.12.

Pulverization (**pvz**) indicates a situation where a cell contains both chromatid and/or chromosome gaps and breaks which are not normally associated with exchanges and are present in such numbers that they cannot be enumerated. Occasionally, one or more chromosomes in a cell are pulverized while the remaining chromosomes are of normal morphology; e.g., pvz(1) is a pulverized chromosome 1.

Premature chromosome condensation (**pcc**) occurs when an interphase nucleus is prematurely induced to enter mitosis. A pcc may involve a G1 or a G2 nucleus. The chromatin of S-phase nuclei undergoing pcc often appears to be pulverized.

The term *premature centromere division* (**pcd**) may be used to describe premature separation of centromeres in metaphase. The pcd may affect one or more chromosomes in a fraction of the cells.

A *marker chromosome* (**mar**) is a structurally rearranged chromosome in which no part can be identified (see Section 9.2.12).

10.2.2 Banded Preparations

When banded preparations allow adequate identification of chromosome segments or chromosome aberrations, the nomenclature system recommended throughout this report can be used. When not, the observations can be described in words.

10.3 Scoring of Aberrations

In the scoring of aberrations, the main types are **chtg, chtb, chte, chrg, chrb, ace, min, r, dic, tr, qr, der**, and **mar**, and reports should, where possible, give the data under these headings. It is recognized, however, that aberrations are frequently grouped to give adequate numbers for statistical analysis or for some other reason. When this is done, it should be indicated how the groupings relate to the aberrations listed above, e.g.:

chromatid aberrations   chtg, chtb, chte
fragments (= deletions)   chrb, ace, min
asymmetric aberrations   ace, dic, r

The data should not be presented as deduced breakages per cell but in such a manner that it is possible to calculate the number of aberrations per cell.

11 Neoplasia

Described below are definitions of terms and recommendations related to abnormalities commonly seen in neoplasia.

11.1 Clones and Clonal Evolution

11.1.1 Definition of a Clone

A clone is defined as a cell population derived from a single progenitor. It is common practice to infer a clonal origin when a number of cells have the same or closely related abnormal chromosome complements. A clone is therefore not necessarily completely homogeneous because subclones may have evolved during the development of the tumor. A clone must have at least two cells with the same aberration if the aberration is a chromosome gain or a structural rearrangement. If the abnormality is loss of a chromosome, the same loss must be present in at least three cells to be accepted as clonal. The term may need to be operationally defined by the author because the criteria for acceptance will depend on, e.g., the number of cells examined, the nature of the aberration involved, the type of culture, and the time cells spend in vitro prior to harvest. In the special situation when in situ preparations are analyzed, the same structural rearrangement or chromosomal gain must be present in at least two metaphase cells from either different primary culture slides, or from well-separated areas or different cell colonies on the same slide. Loss of a chromosome must be detected in at least three such cells. The karyotype designations of different clones and subclones are separated by a **slant line** (/). For order of clone presentation, see Sections 11.1.4 and 11.1.6.

The general rule in tumor cytogenetics is that only the clonal chromosomal abnormalities found in a tumor should be reported. If, for special reasons, nonclonal aberrations are presented, then these must be clearly separated from the clonal abnormalities and should not be part of the description of the tumor karyotype. When the same abnormal clone has been found in an initial and follow-up study, even in a single cell, it should be reported in the karyotype.

46,XX,t(9;22)(q34;q11.2)[1]/46,XX[19]

Similarly, if a single abnormal cell is confirmed by a different method (e.g., FISH), and thus shown to be clonal, it should be reported in the karyotype (see Section 13.4).

46,XX,del(20)(q11.2q13.3)[1]/46,XX[19].nuc ish(D20S108×1)[40/200]

11.1.2 Clone Size

The number of cells that constitute a clone is given in **square brackets** [ ] after the karyotype. When all cells are normal, the number of cells is still specified. In cancer cytogenetics, the clones are written in order of increasing complexity, irrespective of the size of the clone.

46,XX[20]
> A normal female karyotype identified in 20 metaphase cells.

46,XX,t(8;21)(q22;q22)[23]
> A clone with t(8;21) identified in 23 metaphase cells.

46,XY,t(8;21)(q22;q22)[26]/47,XY,t(8;21)(q22;q22),+21[7]
46,XY,t(8;21)(q22;q22)[26]/47,idem,+21[7]
46,XY,t(8;21)(q22;q22)[26]/47,sl,+21[7]
> A clone with 46 chromosomes identified with a t(8;21) as the sole aberration in 26 cells and a sub-clone with 47 chromosomes with the t(8;21) and trisomy 21 in 7 cells. Alternatively, the terms idem or sl may be used to describe subclones (see Section 11.1.4 for details); however, the terms idem and sl must never be intermixed when describing a single tumor sample.

11.1.3 Mainline

The *mainline* (**ml**) is the most frequent chromosome constitution of a tumor cell population. It is a purely quantitative term to describe the largest clone, and does not necessarily indicate the most basic one in terms of progression. In some situations, when two or more clones are of exactly the same size, a tumor may have more than one **ml**. Alternatively, the use of the terms **idem** and **sl** may be used to describe subclones (see Section 11.1.4 for details).

46,XX,t(9;22)(q34;q11.2)[3]/47,XX,+8,t(9;22)(q34;q11.2)[17]
46,XX,t(9;22)(q34;q11.2)[3]/47,sl,+8[17]
46,XX,t(9;22)(q34;q11.2)[3]/47,idem,+8[17]
> The clone with 47 chromosomes represents the mainline, although it has most probably evolved from the clone with 46 chromosomes.

46,XX,der(2)t(2;5)(p23;q35)[10]/47,XX,+2,der(2)t(2;5)[16]
46,XX,der(2)t(2;5)(p23;q35)[10]/47,sl,+2[16]
46,XX,der(2)t(2;5)(p23;q35)[10]/47,idem,+2[16]
> The clone with 47 chromosomes represents the mainline, although it has most probably evolved from the clone with 46 chromosomes.

11.1.4 Stemline, Sideline and Clonal Evolution

Cytogenetically related clones (subclones) are presented, as far as possible, in order of increasing complexity, irrespective of the size of the clone. The *stemline* (**sl**) is the most basic clone of a tumor cell population and is listed first. All additional deviating subclones are termed *sidelines* (**sdl**). To describe the stemlines or sidelines, these abbreviations, or the term **idem** ['idem] (Latin = same), can be used. If more than one sideline is present, these may be referred to as sdl1, sdl2, and so on.

In tumors with subclones the term **idem** can be used, followed by the additional changes in relation to the stemline, which is listed first. Note that idem always refers to the karyotype listed first. This means that in tumors with multiple subclones each clonal change in addition to the first karyotype will have to be repeated. It also means that all plus and minus signs only refer to changes in relation to the stemline karyotype. As an alternative, for more than one sideline, sl and sdl could be used. Note that when two or more stemlines are present, there may also exist two or more sdl1, sdl2 and so on, which will reduce clarity. In such instances idem is preferred.

46,XX,t(9;22)(q34;q11.2)[3]/47,sl,+8[17]/48,sdl1,+9[3]/49,sdl2,+11[12]
46,XX,t(9;22)(q34;q11.2)[3]/47,idem,+8[17]/48,idem,+8,+9[3]/49,idem,+8,+9,
+11[12]

> The clone with 46 chromosomes represents the stemline; the three subclones with 47, 48 and 49 chromosomes are sidelines. In the subclone with 47 chromosomes, the designation sl indicates the presence of the abnormal chromosomes seen in the stemline, i.e. t(9;22)(q34;q11.2) in addition to +8; this subclone is sideline 1 (sdl1). In the subclone with 48 chromosomes (sdl2), the designation sdl1 indicates the presence of the abnormal chromosomes seen in the first sideline, i.e. t(9;22)(q34;q11.2),+8 in addition to +9, and so on. As an alternative, in each subclone the translocation 9;22 is replaced by idem.

46,XX,t(8;21)(q22;q22)[12]/45,sl,–X[19]/46,sdl1,+8[5]/47,sdl2,+9[8]
46,XX,t(8;21)(q22;q22)[12]/45,idem,–X[19]/46,idem,–X,+8[5]/47,idem,–X,
+8,+9[8]

> The clone with t(8;21) as the sole anomaly is the most basic one and hence represents the stemline; the other subclones are listed in order of increasing karyotypic complexity of the aberrations acquired during clonal evolution.

48,XX,t(12;16)(q13;p11.1),...[23]/49,sl,+6[8]/50,sdl1,+7,–8,+9[4]
48,XX,t(12;16)(q13;p11.1),...[23]/49,idem,+6[8]/50,idem,+6,+7,–8,+9[4]

> The subclone with 49 chromosomes has all the abnormalities seen in the stemline plus an extra chromosome 6; the subclone with 50 chromosomes has all sdl1 abnormalities in addition to trisomy 7, monosomy 8, and trisomy 9.

The term **sl** or **sdl times a number** (×2, ×3, etc.) may be used to designate aberrant polyploid clones. Alternatively, the term **idem times a number** (×2, ×3, etc.) may be used to designate aberrant polyploid clones. Additional abnormalities in the polyploid clone may then be indicated using conventional terminology (see Sections 8.1 and 9.1).

46,XY,t(9;22)(q34;q11.2)[3]/92,slx2[5]/93,sdl1,+8[2]
46,XY,t(9;22)(q34;q11.2)[3]/92,idemx2[5]/93,idemx2,+8[2]

> The clone with the t(9;22) is the stemline. Two additional abnormal subclones are identified, one a doubling product or tetraploid subclone of the stemline (sl) and a near-tetraploid (sdl1) subclone with gain of chromosome 8. As an alternative, idem may be used, but all subclones refer back to the stemline.

45,XY,–7[5]/46,sl,+8[6]/46,XY,t(9;22)(q34;q11.2)[3]/92,sl2×2[5]/93,sl2×2,+8[2]
45,XY,–7[5]/46,idem,+8[6]/46,XY,t(9;22)(q34;q11.2)[3]/92,XXYY,t(9;22)×2[5]/
93,XXYY,t(9;22)×2,+8[2]

In tumors with unrelated clones, there may be clonal evolution arising from each unrelated clone. In this instance, the first stemline shows monosomy 7 and is designated sl in the subclone showing trisomy 8. The second stemline shows t(9;22) and is designated sl2 in the subclone showing tetraploidy. Further clonal evolution is found in a sideline showing gain of chromosome 8, but to avoid confusion between sidelines of sl1 and sl2, the use of the term sdl is avoided when referring to a second stemline. The alternative use of idem is listed below for comparison.

48,XX,t(12;16)(q13;p11.1),...[31]/96,sl×2[6]
48,XX,t(12;16)(q13;p11.1),...[31]/96,idem×2[6]

The subclone with 96 chromosomes represents a doubling product of the stemline with 48 chromosomes.

48,XX,t(12;16)(q13;p11.1),...[27]/97,sl×2,+8[3]
48,XY,t(12;16)(q13;p11.1),...[27]/97,idem×2,+8[3]

The subclone with 97 chromosomes represents a doubling product of the hyperdiploid stemline and also has an extra chromosome 8, i.e., there are five chromosomes 8 in this near tetraploid subclone.

48,XX,t(12;16)(q13;p11.1),...[7]/96,sl×2,inv(3)(q21q27),t(3;6)(p25;q21)[19]
48,XX,t(12;16)(q13;p11.1),...[7]/96,idem×2,inv(3)(q21q27),t(3;6)(p25;q21)[19]

The mainline with 96 chromosomes is a doubling product of the hyperdiploid stemline and has in addition an inv(3) and a balanced t(3;6), i.e., there are two normal chromosomes 3 and three normal chromosomes 6 in this near tetraploid subclone.

48,XX,t(12;16)(q13;p11.1),t(14;19)(q23;p11),+17,–19,+20,+21[32]/49,sl,+6[17]
48,XX,t(12;16)(q13;p11.1),t(14;19)(q23;p11),+17,–19,+20,+21[32]/49,idem,+6[17]

The subclone with 49 chromosomes has all the abnormalities seen in the stemline plus an extra chromosome 6.

53,XY,t(14;18)(q32;q21),...[21]/53,sl,del(7)(q21)[9]
53,XY,t(14;18)(q32;q21),...[21]/53,idem,del(7)(q21)[9]

The sideline has a deletion of the long arm of chromosome 7 in addition to the abnormalities seen in the stemline.

53,XY,t(1;6)(p34;q22),–3,...[13]/57,sl,+3,+del(7)(q11.2),+8,+9[22]
53,XY,t(1;6)(p34;q22),–3,...[13]/57,idem,+3,+del(7)(q11.2),+8,+9[22]

There are four additional changes in the subclone with 57 chromosomes in relation to the stemline. Note, however, that the stemline has monosomy 3 whereas the sideline has two normal chromosomes 3, i.e., +3 in this situation does not denote that the clone has trisomy 3.

49,XX,inv(6)(p21q25),...[17]/52,sl,–inv(6),+7,+8,+9,+mar[11]
49,XX,inv(6)(p21q25),...[17]/52,idem,–inv(6),+7,+8,+9,+mar[11]

The inv(6) present in the stemline has not been found in the sideline with 52 chromosomes. The breakpoints in the inv(6) need not be repeated.

11.1.5 Composite Karyotype

In many instances, especially in solid tumors, there is great karyotypic heterogeneity within the tumor, but different cells nevertheless share some cytogenetic characteristics. *Every effort should be made to describe the subclones so that clonal evolution is made evident.* However, in some instances, a *composite karyotype* (**cp**) will have to be created. The composite karyotype contains all clonally occurring abnormalities and should also give the range of chromosome numbers in the metaphase cells containing the clonal abnormalities. The total number of cells in which the clonal changes were observed is given in square brackets after the karyotype, preceded by the abbreviation cp. The term cp should not be used to describe random loss.

47~55,XX,del(3)(p12),+i(6)(p10),del(7)(q11.2),+8,dup(11)(q13q25),+16,+17, der(18)t(18;20)(q23;q11.1),+21,+21,+22[cp24]

> Each of the abnormalities in this example has been seen in at least two cells, but there may be no cell with all abnormalities. The fact that it is a composite karyotype is obvious from the symbol cp and also because the chromosome number is given as a range.

It is not apparent from a composite karyotype how many cells have each abnormality. This information may be expressed by providing the number of cells in square brackets after each abnormality.

45~48,XX,del(3)(p12)[2],–5[4],+8[2],+11[3][cp7]

> In this composite karyotype, constructed on the basis of a total of seven cells, each with at least one of the abnormalities listed, two cells had a terminal deletion of the short arm of chromosome 3 with a breakpoint in 3p12, four cells had monosomy 5, two cells had trisomy 8, and three cells had trisomy 11. Obviously, some cells had more than one of these abnormalities.

It should be noted that in a composite karyotype the sum of the aberrations listed may indicate a higher or lower chromosome number than that actually seen. For example, if the following five cells are karyotyped:

48,XX,+7,+9
48,XX,+7,+11
48,XX,+9,+11
48,XX,+9,+13
48,XX,+13,+21

then the composite karyotype will be:

48,XX,+7,+9,+11,+13[cp5]

> The chromosome number in each of the five cells containing a clonal abnormality is 48, which is thus given as the chromosome number of the composite karyotype, although the sum total of all clonal changes indicates a chromosome number of 50. However, no cell with 50 chromosomes was observed.

Note also that a composite karyotype may contain such seemingly paradoxical abnormalities as loss and gain of the same chromosome. For example, if the following six cells are karyotyped:

45,XX,−15,del(17)(q11.1)
46,XX,+7,−15,del(17)(q11.1)
46,XX,+12,−15
47,XX,+7
47,XX,+15,del(17)(q11.1)
48,XX,+12,+15

then the composite karyotype will be:

45~48,XX,+7,+12,+15,−15,del(17)(q11.1)[cp6]

> Trisomy 15 and monosomy 15 are both clonal changes, present in two and three cells, respectively.

42,XX,−2,−16,−21,−22
44,XX,−1,−7,+8,−11
44,XX,−7,+8,−12,−13
44,XX,−7,−20
46,XX,−7,+8

Composite karyotype:

44~46,XX,−7,+8[cp4]

> Note that the cell with 42 chromosomes is not included because the abnormalities seen are due to random loss and are not part of the clone.

51,XY,+1,−7,+8,t(9;22)(q34;q11.2),+11,+13,+19,+der(22)t(9;22)
51,XY,+1,+5,−7,+8,t(9;22)(q34;q11.2),+11,+19,+der(22)t(9;22)
51,XY,+1,+5,−7,+8,t(9;22)(q34;q11.2),+13,+19,+der(22)t(9;22)
52,XY,+1,+5,−7,+8,t(9;22)(q34;q11.2),+11,+13,+19,+der(22)t(9;22)
46,XY,t(9;22)(q34;q11.2)[5]

Composite karyotype:

46,XY,t(9;22)(q34;q11.2)[5]/51~52,sl,+1,+5,−7,+8,+11,+13,+19,+der(22)t(9;22)
[cp4]
46,XY,t(9;22)(q34;q11.2)[5]/51~52,idem,+1,+5,−7,+8,+11,+13,+19,+der(22)t(9;22)
[cp4]

> This is an example of a stemline with a 9;22 translocation. Multiple cells were found with gain and loss of chromosomes, likely due to clonal evolution. But because not all aberrations are present in all cells, these cells have been combined into a composite karyotype.

11.1.6 Unrelated Clones

Clones with completely unrelated karyotypic anomalies are presented according to their size; the largest first, then the second largest, etc. A normal diploid clone, when present, is always listed last.

46,XX,t(3;9)(p13;p13)[14]/48,XX,+3,+9[11]/46,XX,t(1;6)(p11;p12)[9]/47,XX,t(6;10)(q12;p15),+7[6]/46,XX,inv(6)(p22q23)[3]/46,XX[7]

> Five different clones in the same tumor presented in order of decreasing frequency, irrespective of chromosome number or type of aberration.

If a tumor contains both related and unrelated clones, the former are presented first in order of increasing complexity (see Section 11.1.4), followed by the unrelated clones in order of decreasing frequency.

50,XX,t(2;6)(p22;q16),…[19]/51,sl,+8[7]/52,sdl1,+9[12]/46,XX,del(3)(q13)[11]/47,XX,+7[6]/46,XX,t(1;3)(p22;p14)[4]
50,XX,t(2;6)(p22;q16),…[19]/51,idem,+8[7]/52,idem,+8,+9[12]/46,XX,del(3)(q13)[11]/47,XX,+7[6]/46,XX,t(1;3)(p22;p14)[4]

> The three related clones take precedence over the unrelated clones and are presented first. The three unrelated clones are presented in order of decreasing frequency.

However, if a previously identified abnormality is found among other unrelated clones, it should be listed first, regardless of the number of cells identified.

46,XY,t(9;22)(q34;q11.2)[6]/46,XY,t(1;3)(p22;p14)[14]

11.2 Modal Number

The *modal number* (**mn**) is the most common chromosome number in a tumor cell population. The modal number may be expressed as a range between two chromosome numbers.

Modal numbers in the *haploid* (**n**), *diploid* (**2n**), *triploid* (**3n**) or *tetraploid* (**4n**) range, or near but not equal to any multiple of the haploid number, and which cannot be given as a precise number, may be expressed as *near-haploid* (**n±**), *hypohaploid* (**n–**), *hyperhaploid* (**n+**), *near-diploid* (**2n±**), *hypodiploid* (**2n–**), *hyperdiploid* (**2n+**), *near-triploid* (**3n±**), *hypotriploid* (**3n–**), *hypertriploid* (**3n+**), *near-tetraploid* (**4n±**), *hypotetraploid* (**4n–**), *hypertetraploid* (**4n+**), and so on. Each range is determined as n±n/2, with n/2 defined operationally as 11 chromosomes. Suggested examples of ploidy levels, including ranges of chromosome numbers constituting each level, are given below.

Near-haploidy (23±)	**≤34**	**Near-pentaploidy (115±)**	**104–126**
Hypohaploidy	<23	Hypopentaploidy	104–114
Hyperhaploidy	24–34	Hyperpentaploidy	116–126
Near-diploidy (46±)	**35–57**	**Near-hexaploidy (138±)**	**127–149**
Hypodiploidy	35–45	Hypohexaploidy	127–137
Hyperdiploidy	47–57	Hyperhexaploidy	139–149
Near-triploidy (69±)	**58–80**	**Near-heptaploidy (161±)**	**150–172**
Hypotriploidy	58–68	Hypoheptaploidy	150–160
Hypertriploidy	70–80	Hyperheptaploidy	162–172
Near-tetraploidy (92±)	**81–103**	**Near-octaploidy (184±)**	**173–195**
Hypotetraploidy	81–91	Hypooctaploidy	173–183
Hypertetraploidy	93–103	Hyperoctaploidy	185–195

Pseudodiploid, pseudotriploid, etc., are used to describe a karyotype, which has the number of chromosomes equal to a multiple of the haploid number (*euploid*) but is abnormal because of the presence of acquired numerical and/or structural aberrations. All chromosome numbers deviating from euploidy are *aneuploid*.

The description of sex chromosome abnormalities poses a special problem in male tumors with uneven ploidy levels (haploid, triploid, pentaploid, etc.) because the expected sex chromosome constitution cannot be deduced. For example, the sex chromosome constitution of a triploid tumor might theoretically be XXY or XYY. By convention, in males all sex chromosome deviations should be expressed in relation to X in haploid tumors, to XXY in triploid tumors, to XXXYY in pentaploid tumors, and so on.

11.3   Constitutional Karyotype

The same clonality criteria (see Section 11.1.1) apply to cells containing the constitutional karyotype as to cells containing acquired chromosome abnormalities. A normal diploid clone, when present, is always listed last.

A constitutional anomaly is indicated by the letter **c** after the abnormality designation. In the description of the karyotype, the constitutional anomaly is listed in chromosome number order (see Chapter 6). A clone with only a constitutional anomaly is, as the normal diploid clone, always listed last.

48,XX,+8,+21c[20]
> Tumor cells with a constitutional trisomy 21 and an acquired trisomy 8.

47,X,t(X;18)(p11.1;q11.1),+21c[20]
> Tumor cells with a constitutional trisomy 21 and an acquired t(X;18).

47,XXYc,t(9;22)(q34;q11.2)[20]
> Tumor cells with a constitutional XXY and an acquired t(9;22).

48,XY,+8,inv(10)(p12q22)c,+21[20]
> Tumor cells with a constitutional inv(10) and acquired trisomies 8 and 21.

47,XX,del(5)(q15),+mar c[20]
> Tumor cells with a constitutional marker chromosome of unknown origin and an acquired deletion of the long arm of one chromosome 5. For constitutional markers, there is a space between **mar** and **c**.

48,XY,+8,+21c[3]/49,idem,+9[5]/47,XY,+21c[12]
> Tumor cells with a constitutional trisomy 21 and acquired trisomies 8 and 9. The clone with only the constitutional trisomy 21 is listed last irrespective of the size of this clone.

49,XX,t(2;13)(q37;q14),+18c,+18,+mar[3]/47,XX,+18c[17]
> Tumor cells with a constitutional trisomy 18 as the sole anomaly in one clone and with acquired abnormalities, including an additional chromosome 18, in another clone. The clone with 49 chromosomes thus has four chromosomes 18. The clone with only the constitutional trisomy 18 is listed last.

To appropriately express acquired abnormalities affecting one of the chromosomes of a pair that is involved in a constitutional anomaly, the constitutional aberration must always be given, even if none of the tumor cells have this particular aberration. Thus, an acquired abnormality is always presented in relation to the constitutional karyotype.

46,XX,+21c,–21[20]

> The patient has a constitutional trisomy 21 and the acquired abnormality in the tumor cells is a loss of one chromosome 21.

45,Xc,t(X;18)(p11.1;q11.1)[20]

> Tumor cells in a patient with Turner syndrome (45,X) have an acquired t(X;18), i.e., the only X chromosome is involved in the translocation and consequently there is no normal X chromosome in the tumor cells.

46,XX,der(9)t(9;11)(p22;q23)t(11;12)(p13;q22)c,der(11)t(9;11)t(11;12)c,der(12) t(11;12)c[20]

> Female patient with a known constitutional t(11;12)(p13;q22) presents with t(9;11) positive AML. The derivative chromosome 11 involved in the t(11;12)c is also involved in the t(9;11) aberration. The resulting karyotype, with both constitutional and acquired aberrations, should list each aberrant chromosome as a derivative chromosome.

12 Meiotic Chromosomes

During late prophase-first metaphase, the bivalents may be grouped by size, and bivalent 9 can sometimes be distinguished by its secondary constriction. At these stages, the Q- and C-staining methods are particularly informative. The autosomal bivalents generally show the same Q-band patterns as somatic chromosomes. The C-staining method reveals the centromere position, thus allowing identification of the bivalents in accordance with the conventionally stained somatic chromosomes. There are, however, minor differences in the C-band patterns between the bivalents and mitotic chromosomes.

When the Q- and C-staining methods are used consecutively, further distinction of the bivalents is possible. Measurements of the relative length of orcein-stained bivalents, previously identified by these special techniques, are in good agreement with corresponding mitotic measurements. Chiasma frequencies have been determined for individual bivalents.

The Y chromosome can be identified at all meiotic stages by the intense fluorescence of its long arm. Both the Q- and C-staining methods have revealed that the short arm of the Y is associated with the short arm of the X in the first meiotic metaphase.

12.1 Terminology

The abbreviations **PI**, **MI**, **AI**, **MII**, and **AII** are used to indicate the stage of meiosis, namely, *prophase of the first division, first metaphase* (including diakinesis), *first anaphase, second metaphase*, and *second anaphase*. This is followed by the total count of separate chromosomal elements. The sex chromosomes are then indicated by XY or XX when associated and as X,Y when separate. Any additional, missing, or abnormal element follows, with that element specified within parentheses and preceded by the Roman numeral **I**, **II**, **III**, or **IV** to indicate if it is a *univalent, bivalent, trivalent*, or *quadrivalent*, respectively. The absence of a particular element is indicated by a **minus** (–) sign. The **plus** (+) sign is used in first metaphase only when the additional chromosome is not included in a multivalent. The chromosomes involved in a rearrangement are listed numerically within parentheses and separated by a **semicolon** (;).

A more detailed description, for instance, of the chromosomal segments involved in a rearrangement may be included within parentheses using the standard nomenclature, with which this meiotic notation has been designed to conform. When necessary, use of the abbreviations **fem** and **mal** is recommended for *female* and *male*, respectively, and when a more detailed description of different premeiotic and meiotic stages is required, the following abbreviations may be used:

spm	Spermatogonial metaphase
oom	Oogonial metaphase
lep	Leptotene
zyg	Zygotene
pac	Pachytene
dip	Diplotene
dit	Dictyotene
dia	Diakinesis

The abbreviation **xma** is suggested for *chiasma*(*ta*). The total number of chiasmata in a cell can be designated by placing this abbreviation, followed by an **equal sign** (=) and a two-digit number, in parentheses, e.g., (xma=52). In the case of a meiotic cell with a low number of chiasmata, a single digit should be preceded by a zero, e.g., (xma=09).

The number of chiasmata in a bivalent or multivalent or their arms may be indicated by a single digit, e.g., (xma=4).

Location of chiasmata can be indicated by the standard arm symbols **p** and **q**, supplemented by **prx** for *proximal*, **med** for *medial*, **dis** for *distal*, and **ter** for *terminal*. The band or region number can be used when such precise information is available.

Chromosomes participating in a bivalent or multivalent are specified within parentheses after the Roman numeral that describes the bivalent (II) or the type of multivalent (III, IV, etc.). If the number of chiasmata within the multivalent is known, this is indicated within parentheses in consecutive order, i.e., the number of chiasmata between the first and second chromosome is given first, between the second and the third next, etc. The last figure then indicates the number of chiasmata between the last and first chromosome. If the number of chiasmata in non-interstitial and interstitial segments can be specified separately, these should be represented by a **plus** (+) sign. The number of chiasmata in the non-interstitial segment is written first, e.g., (xma=2+1), indicating two chiasmata in the non-interstitial and one in the interstitial segment. It is assumed that a careful description of the mitotic karyotype of the subject will be given separately.

12.1.1   Examples of Meiotic Nomenclature

MI,23,XY

  A primary spermatocyte at diakinesis or metaphase I with 23 elements, including an XY bivalent.

MI,24,X,Y

  A primary spermatocyte at diakinesis or metaphase I with 24 elements, including X and Y univalents.

MI,23,XY,III(21)

  A primary spermatocyte with 23 elements from a male with trisomy 21. The three chromosomes 21 are represented by a trivalent.

MI,24,XY,+I(21)

A primary spermatocyte with 24 elements from a male with trisomy 21. The extra chromosome 21 is represented by a univalent.

MI,22,XY,III(13q14q)

A primary spermatocyte with 22 elements from a der(13;14)(q10;q10) heterozygote. The translocation chromosome is represented by a trivalent.

MI,23,XY,(xma=52)

Spermatocyte in first metaphase with 23 elements, including an XY bivalent. The total number of chiasmata in the cell is 52, the association between the X and Y chromosomes being counted as one chiasma.

fem dia,II(2,2)(xma=4)

Oocyte in diakinesis in which bivalent 2 has four chiasmata.

fem dia,II(2,2)(xma=4)(p=2,q=2)

Female diakinesis in which bivalent 2 has four chiasmata. The positions of the chiasmata are known. Thus (xma=4)(p=2,q=2) indicates that there are two chiasmata on the short arm and two on the long arm. More precise location of the chiasmata could then be indicated, e.g., by (xma=4) (pter,pprx,qmed,qdis). Alternatively, if the chiasmata have been localized to specific regions, these could be indicated, e.g., by (xma=4)(pter,p1,q2,qter).

mal MI,III(14,14q21q,21)(xma=3)

Male first metaphase with a trivalent composed of one chromosome 14, one 14q21q Robertsonian translocation chromosome, and one chromosome 21. There are three chiasmata, the positions of which have not been specified.

MI,23,X,Y,III(13,13q14q,14)(xma=2,1),(xma=51)

Spermatocyte in first metaphase with 23 elements, univalent X and Y chromosomes, and one trivalent composed of one chromosome 13, one 13q14q Robertsonian translocation chromosome, and one chromosome 14. There are two chiasmata between the normal chromosome 13 and the 13q14q translocation chromosome and one chiasma between the translocation chromosome and the normal chromosome 14. Altogether, there are 51 chiasmata in the cell.

fem dia,IV(2,der(2),5,der(5))(xma=2+1,1,1+0,1)

Oocyte in diakinesis with a quadrivalent composed of two normal chromosomes and two derivative chromosomes of chromosomes 2 and 5, respectively. There are three chiasmata between chromosomes 2 and der(2), of which two are in the non-interstitial segment and one is in the interstitial segment. In addition, there is one chiasma between der(2) and chromosome 5, one in the non-interstitial and none in the interstitial segment between chromosome 5 and der(5), and finally one between der(5) and chromosome 2. The last chiasma indicates that the quadrivalent has a ring shape.

MI,24,X,Y,III(2,der(2),5)(xma=4),I(der(5)),(xma=51)

Spermatocyte in first metaphase with 24 elements, including univalent X and Y chromosomes, one trivalent, and one additional univalent. The trivalent is composed of one normal and one derivative chromosome 2, as well as one normal chromosome 5. This trivalent has a total of four chiasmata, the positions of which are not known. One univalent is composed of one derivative chromosome 5. The total number of chiasmata in the cell is 51.

MII,22,X,–16,+16cht,+16cht

Oocyte at second meiotic metaphase in which chromosome 16 is absent, but is represented by its two single chromatids.

12.1.2 Correlation between Meiotic Chromosomes and Mitotic Banding Patterns

Meiotic chromosomes from pachytene spermatocytes have been shown to exhibit chromomere patterns without any pretreatment, which correspond well with Giemsa-dark bands of somatic chromosomes, suggesting that both represent a basic structural feature of the mammalian chromosome. Just as in the case of somatic chromosome bands, the number of chromomeres that can be recognized is a function of the stage of contraction. The chromomere patterns of human oocyte pachytene chromosomes are apparently similar to those of spermatocyte chromosomes, although the former exhibit less contraction and hence more chromomeres.

Figure 9 is an idiogram of the 22 autosomal bivalents from pachytene spermatocytes employing the nomenclature of the somatic chromosome banding patterns using the 850-band stage nomenclature because this stage of contraction corresponds approximately to mid-pachytene of spermatocyte meiosis. The chromomeres are given the numbers of Giemsa-positive bands and inter-chromomere regions are given the numbers of Giemsa-negative bands. A special feature of human pachytene chromosomes is the presence of a particulate or puff-like structure located at the heterochromatic region (9q12). The structure is transient and is limited to the pachytene stage.

Fig. 9. Chromomere idiogram of the 22 autosomal bivalents at pachytene. The nomenclature used is that of the 850-band stage of somatic chromosomes (ISCN 1981). Chromomere numbers are equivalent to those of Giemsa-positive bands of somatic chromosomes and inter-chromomere region numbers are equivalent to those of Giemsa-negative bands of somatic chromosomes. Individual bivalent idiograms compared with photomicrographs of bivalents (two Giemsa-stained and one quinacrine-stained). Lines between idiograms and photomicrographs connect centromeres and chromomeres that correspond to landmark bands of somatic chromosomes. (Method of Jhanwar et al., 1982; courtesy of Drs. R.S.K. Chaganti and S.C. Jhanwar).

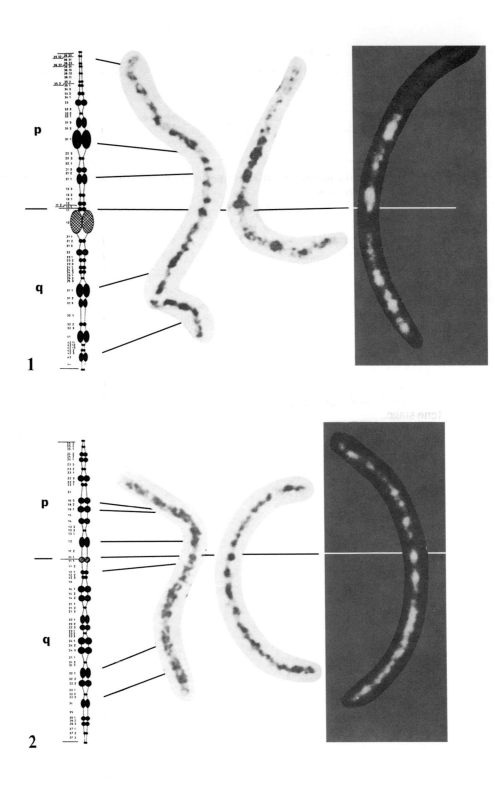

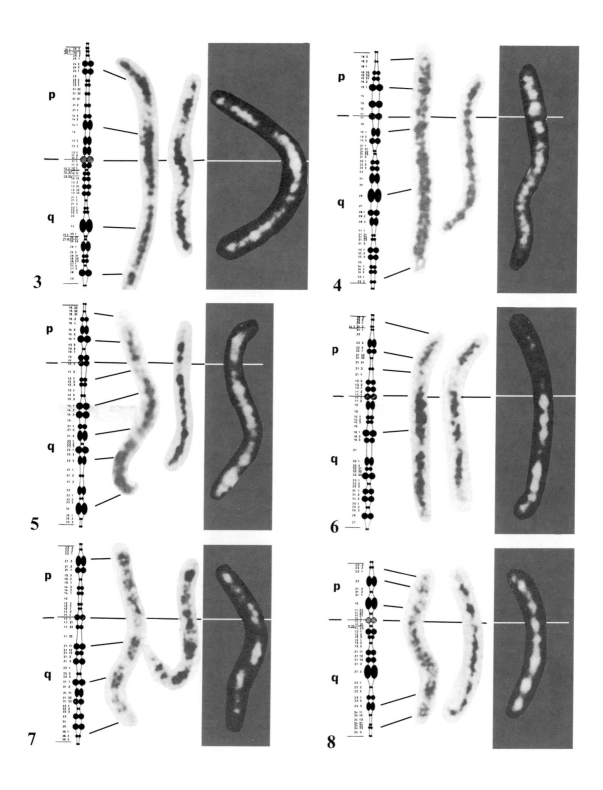

Fig. 9 continued (see legend on p 100)

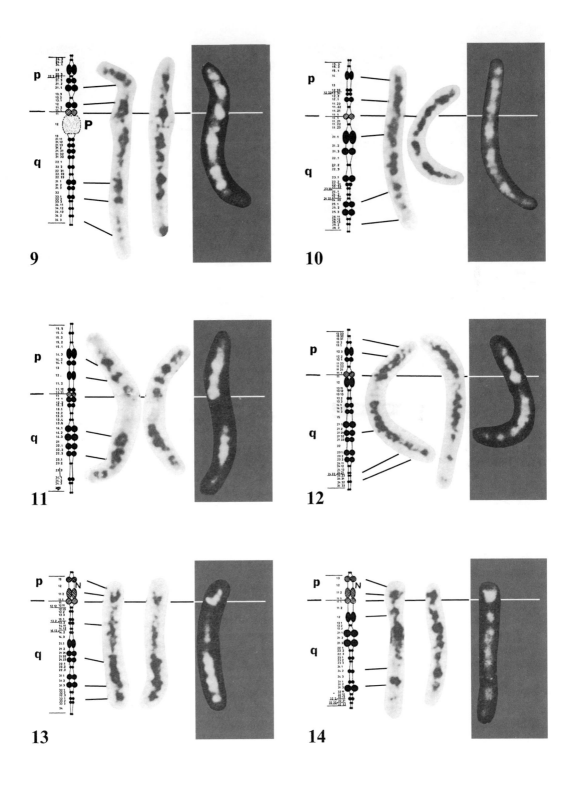

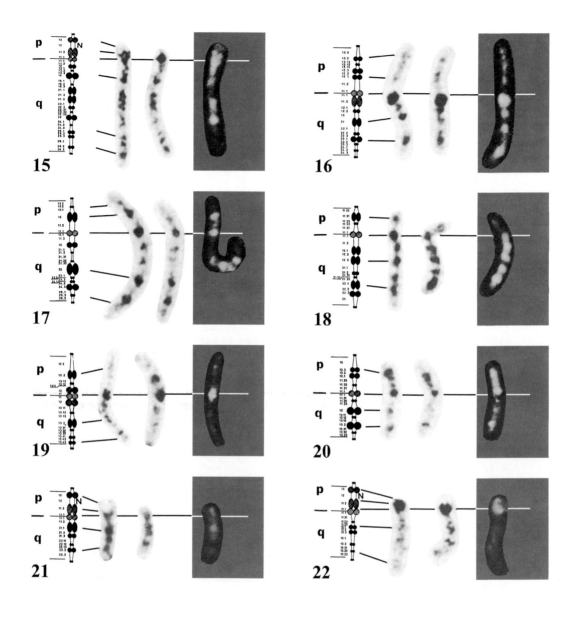

Fig. 9 continued (see legend on p 100)

13 In situ Hybridization

13.1 Introduction

A major advance in human cytogenetics since the publication of ISCN (1985) has been the development and implementation of a variety of non-isotopic in situ hybridization techniques to detect (Lichter et al., 1990; Trask, 1991), and in some instances quantify (Kallioniemi et al., 1992), specific DNA sequences and to locate them to specific chromosomal sites. The ever-increasing availability of a number of sequence-specific DNA probes, and their amplification by the polymerase chain reaction and the availability of fluorochrome-tagged reporter molecules that are bound to DNA probes, have all contributed to bridging the gap between the microscope and the molecule.

Techniques utilizing fluorescence in situ hybridization (FISH) allow the use of a number of fluorochromes so that the locations of different and differently tagged probes, and the relative positions of their binding sites, may be visualized microscopically on a single chromosome segment or DNA/chromatin fiber (Wiegant et al., 1993). In addition, the use of composite probes, coupled with suppressive hybridization (Landegent et al., 1987), enables whole chromosomes, or chromosome segments, to be specifically 'painted' and uniquely visualized (Lichter et al., 1988; Pinkel et al., 1988; Guan et al., 1994). FISH banding methods are available and reviewed in Liehr et al. (2006). These developments have also enabled the cytogeneticist to detect the presence of specific DNA sequences in interphase nuclei and to visualize their distribution (Cremer et al., 1986). FISH applied to free, linearly extended chromatin fibers or naked DNA strands has increased the resolution of FISH interphase mapping to ≤ 1 kb (Wiegant et al., 1992; Parra and Windle, 1993).

FISH techniques have provided the cytogeneticist with an increased ability to identify chromosome segments, to correlate chromosome structures with gene locations, to reveal cryptic abnormalities that are undetectable using standard banding techniques, and to analyze and describe complex rearrangements. If FISH further clarifies the karyotype and, in retrospect, the abnormality can be visualized with banding, the karyotype may be re-written to reflect this new FISH information. If the abnormality is cryptic and cannot be visualized by banding, the abnormality should not be listed in the banded karyotype.

13.2 List of Symbols and Abbreviations

A list of symbols and abbreviations is also found in Chapter 3.

- –   Absent from a specific chromosome
- +   Present on a specific chromosome
- ++   Two hybridization signals or hybridization regions on a specific chromosome

×	Multiplication sign, precedes the number of signals seen
.	Period, separates cytogenetic observations from results of in situ hybridization
;	Semicolon, separates probes on different derivative chromosomes
amp	Amplified signal
cgh	Comparative genomic hybridization
con	Connected signals (signals are adjacent)
dim	Diminished signal intensity
enh	Enhanced signal intensity
fib ish	Extended chromatin/DNA fiber in situ hybridization
ish	In situ hybridization; when used without a prefix applies to chromosomes (usually metaphase or prometaphase) of dividing cells
nuc ish	Nuclear or interphase in situ hybridization
pcp	Partial chromosome paint (hybridization with probe mixtures prepared from partial chromosome scrapings, contigs, etc.)
rev ish	Reverse in situ hybridization including comparative genomic in situ hybridization
sep	Separated signals (signals are separated)
subtel	Subtelomeric
wcp	Whole chromosome paint

13.3 Prophase/Metaphase in situ Hybridization (ish)

If a standard cytogenetic observation has been made, it may be given followed by a **period** (.), the abbreviation **ish**, a space, and the ish results. If a standard cytogenetic observation has not been made, the ish observations only are given.

Observations on structurally abnormal chromosomes are expressed by the symbol **ish**, followed by a space and then the symbol for the structural abnormality (whether seen by standard techniques and ish or only by ish), followed in separate parentheses by the chromosome(s), the breakpoint(s), and the locus or loci for which probes were used. When available, the clone name is preferred. If the clone name is not available, the locus designated according to GDB (Genome Database) should be used. If no GDB locus is available, the gene name can be used, according to the HUGO-approved nomenclature. Although gene acronyms are usually italicized, they should not be italicized in the nomenclature. Thus, at the discretion of the investigator or laboratory director the clone name or accession number, gene name, GDB D-number or database nucleotide number in a specified genome build [e.g. NCBI Build 35 (B35)] can be used:

ish del(22)(q11.2q11.2)(N25–)
ish del(22)(q11.2q11.2)(HIRA–)
ish del(22)(q11.2q11.2)(D22S75–)
ish del(22)(q11.2q11.2)(B35:CHR22:12345678–12345778–)

The locus designations (in capital letters but not in italics) are separated by a comma and the status of each locus is given immediately after the locus designation, e.g., **present** (+) or **absent** (–), for example:

46,XY.ish del(22)(q11.2q11.2)(D22S75–)

Observations on normal chromosomes are expressed by the symbol ish followed by a space and the chromosome, region, band, or sub-band designation of the locus or loci tested (not in parentheses), followed in parentheses by the locus (loci) tested, a **multiplication sign** (×) and the number of signals seen, for example:

46,XY.ish 22q11.2(D22S75×2)

> Conventional cytogenetic analysis showed a normal male karyotype and FISH using a probe in the DiGeorge syndrome region (D22S75) showed a normal hybridization pattern on metaphase chromosomes.

46,XX.ish 22q11.2(D22S75×2)

> A female with a normal karyotype by cytogenetic analysis is normal by ish using a probe for locus D22S75.

46,XX.ish del(22)(q11.2q11.2)(D22S75–)

> A female with a normal karyotype by cytogenetic analysis has a deletion in the DiGeorge syndrome critical region (DGCR) on chromosome 22 identified by ish using a probe for locus D22S75.

ish del(22)(q11.2q11.2)(D22S75–)

> Conventional cytogenetic analysis was not performed but a deletion in the DGCR on chromosome 22 was identified by ish using a probe for locus D22S75.

ish del(22)(q11.2q11.2)(D22S75–),del(22)(q11.2q11.2)(D22S75–)

> Conventional cytogenetic analysis was not performed but a deletion in the DGCR in both chromosomes 22 was identified by ish using a probe for locus D22S75.

ish del(22)(q13.3q13.3)(ARSA–)

> Conventional cytogenetic analysis was not performed but a deletion of distal 22q was identified by ish using a probe to the *ARSA* locus.

ish 22q11.2(HIRA×2),del(22)(q13.3)(ARSA–)

> Only the clinically relevant or informative results need to be in the karyotype. Both examples describe a deletion of the *ARSA* locus after using a cocktail containing the *ARSA* and *HIRA* loci. Either example is correct.

46,XX.ish del(7)(q11.23q11.23)(ELN–)

> A microdeletion in the Williams syndrome region of chromosome 7 identified by ish with an elastin gene (*ELN*) probe.

46,XX,del(15)(q11.2q13).ish del(15)(q11.2q11.2)(SNRPN–,D15S10–)

> A cytogenetically detected deletion of bands 15q11.2q13 characterized by ish. Two loci (*SNRPN* and D15S10) from the Prader-Willi/Angelman region are deleted.

46,XY.ish del(15)(q11.2q11.2)(SNRPN–,D15S10–)

> A microdeletion of the Prader-Willi/Angelman region of chromosome 15 identified by ish. The deletion includes the region defined by probes for the *SNRPN* and D15S10 loci.

46,XY.ish del(15)(q11.2q12)(D15S11+,SNRPN−,D15S10−,GABRB3+)

A microdeletion of chromosome 15 defined by ish using probes for loci D15S11, *SNRPN*, D15S10 and *GABRB3*. *SNRPN* and D15S10 are deleted while D15S11 and *GABRB3* are retained.

46,XY.ish dup(17)(p11.2p11.2)(RAI1++)

The region containing the *RAI1* locus on chromosome 17 is duplicated as detected by ish on metaphase chromosomes. There is one signal in the homologous chromosome 17, not indicated in the karyotype.

46,XY.ish 17p11.2(RAI1×2)

The *RAI1* locus on chromosome 17 is present in the normal copy number (two copies) as determined by ish with a locus-specific probe.

46,XX,add(4)(q35).ish dup(4)(q33q35)(wcp4+)

A chromosome 4 has extra material attached at band 4q35. Utilizing a whole chromosome paint 4, the extra chromatin was identified as a duplicated region of chromosome 4, determined by G-banding to be 4q33q35.

46,XX,add(4)(q31).ish der(4)dup(4)(q31q34)(wcp4+)add(4)(q34)(wcp4−)

A chromosome 4 has extra chromatin attached at band 4q31. Using whole chromosome paint 4 the proximal part of the additional material was shown to be derived from chromosome 4. G-banding suggested a duplication of bands 4q31 to 4q34. However, there was additional material distal to the duplication which did not hybridize with whole chromosome 4 paint, and is therefore of unknown origin.

46,X,r(X).ish r(X)(p22.3q21)(KAL+,DXZ1+,XIST+,DXZ4−)

A ring X was further defined by ish as containing the short arm marker *KAL1*, the X alpha-satellite DXZ1 and the *XIST* gene on the long arm. It does not include DXZ4 at Xq24.

46,X,+r.ish r(X)(wcpX+,DXZ1+)

A ring chromosome was identified by ish as a derivative X chromosome using whole chromosome paint X and X alpha-satellite probe DXZ1.

46,X,?i(Y)(p10).ish idic(Y)(q11)(DYZ3++,DYZ1−)

A presumed isochromosome for the short arm of Y was shown by ish to have two centromeres and no heterochromatin.

46,XX.ish der(X)t(X;Y)(p22.3;p11.3)(SRY+)

A presumed unbalanced translocation between the X and Y short arms resulting in a derivative X containing *SRY* at Xp22.3.

45,XY,der(14;21)(q10;q10).ish dic(14;21)(p11.2;p11.2)(D14Z1/D22Z1+;D13Z1/D21Z1+)

A Robertsonian translocation, rob(14;21) or der(14;21), shown to be dicentric using ish.

45,XY,der(14;21)(p11.2;p11.2).ish dic(14;21)(p11.2p11.2)(D14Z1/D22Z1+;D13Z1/D21Z1+)

A Robertsonian translocation shown to be dicentric using ish. Therefore, the G-banded interpretation has been re-written, as compared to the previous example, to reflect the FISH results and the reinterpretation of the breakpoints.

46,XX.ish t(4;11)(p16.3;p15)(wcp11+,D4F26–,D4S96+,D4Z1+;D4F26+,wcp11+)

A cryptic reciprocal translocation between chromosomes 4 and 11 was identified by ish. The der(4) was positive with whole chromosome paint 11, a probe for D4S96 (Wolf-Hirschhorn region) and 4 alpha-satellite but negative for D4F26 (4p telomere region). The der(11) was positive for D4F26 as well as whole chromosome paint 11.

46,XX.ish der(4)t(4;11)(p16.3;p15)mat(wcp11+,D4F26–,D4S96+,D4Z1+)

This child is an unbalanced offspring from the segregation of the cryptic translocation above. She has one normal chromosome 4 and two normal chromosomes 11. The ish results of the der(4) are the same as der(4) of the mother.

46,XY.ish 4p16.3(D4F26,D4S96)×2

A normal male (father of the child in the previous example) was tested by ish using probes for loci D4F26 and D4S96. There were two copies of both.

46,XX,ins(2)(p13q21q31).ish ins(2)(wcp2+)

A direct insertion of the long arm segment 2q21q31 into the short arm at band 2p13 was confirmed as derived from chromosome 2 by ish using whole chromosome paint 2.

46,XY,ins(5;2)(p14;q32q22).ish ins(5;2)(wcp2+)

An inverted insertion of a chromosome 2 segment into the short arm of chromosome 5 was confirmed as derived from chromosome 2 using whole chromosome paint 2.

46,XY,t(9;22)(q34;q11.2)[20].ish t(9;22)(ABL1–;BCR+,ABL1+)[20]

A male karyotype with a t(9;22) that has been characterized by ish using a single-fusion probe. The probe sequence from the *ABL1* locus is missing from the derivative chromosome 9 and is present on the derivative chromosome 22 distal to the *BCR* locus.

47,XY,t(9;22)(q34;q11.2),+der(22)t(9;22)[20].ish t(9;22)(ABL1–;BCR+,ABL1+), der(22)(BCR+,ABL1+)[20]

A male karyotype with a t(9;22) plus an extra copy of the der(22) that has been characterized by ish using a single-fusion probe. The *ABL1* locus is missing from the derivative chromosome 9 and is present on both derivative chromosomes 22 distal to the *BCR* locus.

46,XX,t(9;22)(q34;q11.2)[20].ish t(9;22)(ABL1+,BCR+;BCR+,ABL1+)[20]

A female karyotype with a t(9;22) detected using dual-fusion probes for *BCR* and *ABL1*. One copy of *ABL1* and one copy of *BCR* are found on each derivative chromosome.

46,XX,t(9;22)(q34;q11.2)[20].ish der(9)t(9;22)del(9)(q34q34)(ABL1–,BCR+), der(22)t(9;22)(BCR+,ABL1+)[20]

A female karyotype with a t(9;22) detected using dual-fusion probes for *BCR* and *ABL1*. There is a deletion on the derivative 9, encompassing the *ABL1* locus, not detected using conventional cytogenetic analysis.

46,XX,t(9;22)(q34;q11.2)[20].ish der(9)t(9;22)del(9)(q34q34)(ASS1–,ABL1–,BCR+), der(22)t(9;22)(BCR+,ABL1+)[20]

A female karyotype with a t(9;22) detected using a three-color dual-fusion probe for *ASS1*, *ABL1* and *BCR*. There is a deletion on the derivative 9, encompassing the *ASS1* and *ABL1* loci, not detected using conventional cytogenetic analysis.

46,XX,inv(16)(p13.1q22)[20].ish inv(16)(p13.1)(3'CBFB+)(q22)(5'CBFB+)[20]

Inversion of chromosome 16 separates the two probes for the *CBFB* locus into the 3' probe on the short arm and the 5' on the long arm. Note in this and the following examples that the inversion breakpoints are in separate parentheses to make the FISH information apparent.

46,XX,inv(16)(p13.1q22)[20].ish inv(16)(p13.1)(RP11-620P11+)(q22)(RP11-620P11+)[20]

Inversion of chromosome 16 separates the region corresponding to BAC probe RP11-620P11 giving a signal on the short arm and on the long arm.

46,XX,t(16;16)(p13.1;q22)[20].ish t(16;16)(3'CBFB+;3'CBFB−)[20]

Translocation disrupts the *CBFB* locus resulting in translocation of the 3' probe from 16q22 to 16p13.1 on the other homologue.

46,XX,t(2;17)(q32;q24)[20].ish t(2;17)(AC005181+;AC005181+)[20]

Translocation disrupts the region corresponding to BAC clone AC005181, at 17q24, resulting in a signal on both derivative chromosomes.

46,XX,t(2;17)(q32;q24)[20].ish t(2;17)(CTD-3115L8+;RP11-959M22+)[20]

Same translocation as above. The breakpoint has been located between BAC clone RP11-959M22 and CTD-3115L8 on chromosome 17, which results in CTD-3115L8 moving to chromosome 2 and RP11-959M22 being retained on chromosome 17.

47,XY,+mar.ish der(8)(D8Z1+)

An extra marker chromosome identified by ish to be derived from chromosome 8 using an 8-specific alpha-satellite probe.

47,XY,+mar.ish der(17)(wcp17+,D17Z1+)

An extra marker chromosome identified by ish as derived from chromosome 17 using whole chromosome paint 17 and a 17-specific alpha-satellite probe.

47,XX,+mar.ish der(18)t(18;19)(wcp18+,D18Z1+,wcp19+)

An extra marker chromosome identified by ish to be derived in part from chromosome 18, using whole chromosome paint 18 and an 18 alpha-satellite probe, and from chromosome 19 using whole chromosome paint 19.

47,XY,+mar.ish add(17)(p12)(wcp17+,CMT1A+,D17Z1+)

An extra marker chromosome identified by ish to be partially derived from chromosome 17 using whole chromosome paint 17 and probes for *CMT1A* and D17Z1. Additional material of unknown origin has replaced the segment distal to 17p12.

47,XX,+mar.ish add(16)(p or q)(wcp16+,D16Z1+)

An extra marker chromosome identified by ish to be partially derived from chromosome 16 (arm unknown) using whole chromosome paint 16 and the 16 alpha-satellite probe. Additional material of unknown origin is present in the marker.

mos 46,X,+r[15]/45,X[10].ish r(X)(wcpX+,DXZ1+)

A female with two cell lines, one 45,X and another with 46 chromosomes including a ring. Using ish, the ring was identified as an X using the whole chromosome paint X and the X alpha-satellite probe. The number of cells, not percentage, is given in square brackets.

46,X,+r.ish r(X)(wcpX+,DXZ1+)[15]/r(X)(wcpX+,DXZ1++)[10]

A ring chromosome replacing a sex chromosome was identified by ish as X using whole chromosome paint X. Probe DXZ1, for the X alpha-satellite, showed the ring to be monocentric in some cells and dicentric in other cells.

ish dmin(MYCN×20~50)[20]

Double minutes, identified to contain *MYCN*, are found in 20–50 copies per cell.

13.3.1 Use of dim and enh

46,Y,del(X)(p11.2p11.4).ish del(X)(p11.2p11.4)(RP1-112K5 dim,RP11-265P11 dim)

Deletion of the short arm of chromosome X. On the deleted chromosome, the signals of clones RP1-112K5 (Xp11.2) and RP11-265P11 (Xp11.4) are consistently less intense than on the normal homologue, indicating that they are partially deleted and recognize the proximal and distal breakpoints respectively.

46,XX.ish 17p11.2(RAI1 enh)

Enhanced signal at 17p11.2 using a probe to the *RAI1* locus.

13.3.2 Subtelomeric Metaphase in situ Hybridization

Subtelomeric FISH is usually performed in panels so that the 41 unique chromosome ends are hybridized simultaneously. A short system is appropriate to describe a normal result after using a subtelomeric FISH panel, for example:

ish subtel(41×2)

Normal result using 41 probes to the 41 subtelomeric regions.

ish t(13;20)(q34–,p13+;p13–,q34+)(RP11-63L17–,RP5-1103G7+;RP5-1103G7–, RP11-63L17+)

A balanced translocation between the distal long arm of chromosome 13 and the distal short arm of chromosome 20.

ish der(13)t(13;20)(q34–,p13+)(RP11-63L17–,RP5-1103G7+)

An unbalanced translocation between the distal long arm of chromosome 13 and the distal short arm of chromosome 20. The subtelomeric region of 13q is deleted and replaced with the subtelomeric region of 20p. The designation pter and qter may be used instead of the distal bands. Alternatively, clone names may be used.

13.4 Interphase/Nuclear in situ Hybridization (nuc ish)

Information of interest in *interphase ish*, signified by the symbol **nuc ish**, includes the number of signals and their positions relative to each other. ISCN (1995) provides for the use of a band designation in interphase FISH. This is considered an optional detailed form to be used at the discretion of the investigator or laboratory director. A short nomenclature description has now been provided that does not indicate chromosome band locations, providing for the ambiguity of hybridization location in interphase nuclei in the absence of discernible bands (chromosomes). Especially in the case of amplification, the short nomenclature description is recommended. If a collection of contiguous probes is used to designate a locus, a single designation is used in the nomenclature and the composition is described in the report.

13.4.1 Number of Signals

To indicate the number of signals, the abbreviation **nuc ish** is followed immediately, in parentheses, by the locus designation, a **multiplication sign** (×), and the number of signals seen. If the detailed system is used, a space should follow **ish**, and then the band designation.

   If probes for two or more loci are used in the same hybridization, they follow one another in a single set of parentheses, separated by a **comma (,)**, and a **multiplication sign** (×) outside the parentheses if the number of signals for each probe are the same and inside the parentheses if the number of hybridization signals varies. If multiple probes on the same chromosome are used, they are listed pter→qter, separated by commas. If loci on two different chromosomes are tested, results are reported in a string, separated by commas, in the order sex chromosomes and autosomes 1 to 22. If the study is on a cancer specimen, the number of cells scored is placed in square brackets.

nuc ish(D21S65×2) nuc ish 21q22(D21S65×2)	two copies of locus D21S65
nuc ish(D21S65×3) nuc ish 21q22(D21S65×3)	three copies of locus D21S65
nuc ish(DXZ1×3)	three copies of locus DXZ1
nuc ish(MYCN×12~>50)[200]	twelve to more than 50 copies of *MYCN* found in 200 cells
nuc ish amp(MYCN)[200]	number of *MYCN* copies cannot be quantified because it is increased in copy number beyond that which can be reliably counted
nuc ish(D17Z1,ERBB2)×2[100]	two copies of *ERBB2 (HER-2)* were found in 100 cells with two copies of the centromere 17 probe D17Z1
nuc ish(D17Z1,ERBB2)×4~5[100/200]	four to five copies of D17Z1 and *ERBB2* were found in 100 out of 200 cells

nuc ish(D17Z1×2,ERBB2×10~20)[100/200]	10 to 20 copies of *ERBB2* were found in 100 cells as compared to two copies of the centromere 17 probe D17Z1
nuc ish(D17Z1×2~3),amp(ERBB2) [100/200]/(D17Z1,ERBB2)×3[20/200]	amplification of *ERBB2* was found in comparison to two to three copies of D17Z1 in 100 cells. In addition, there were three copies of both D17Z1 and *ERBB2* in 20 cells.
nuc ish(D21S65,D21S64)×3	three copies of locus D21S65 and three copies of locus D21S64
nuc ish(DXZ1×2,DYZ3×1,D18Z1×3), (RB1,D21S341)×3	three copies of 13, 18 and 21, two copies of X and one copy of Y were found, which may indicate a triploid 69,XXY
nuc ish(ABL1,BCR)×2[400] nuc ish 9q34(ABL1×2),22q11.2(BCR×2)[400]	two copies of each locus *ABL1* and *BCR* found in 400 cells, expressed with or without band designations
nuc ish(KAL1,D21S65)×2	two copies of locus *KAL1* and two copies of locus D21S65
nuc ish(KAL1,GK,DMD)×1	one copy of each locus, listed pter→qter

If chromosome analysis and interphase FISH are performed, each is reported with the string, separated by a period [.].

46,XY[20].nuc ish(TP53×2)[400]

If metaphase and interphase FISH are both performed, each is reported within the string, separated by a period [.]. For cancer studies, the number of cells scored is shown in square brackets for each:

46,XY[20].ish 9q34(ABL1×2),22q11.2(BCR×2)[20].nuc ish(TP53×2)[400]

Interphase analysis may be used to determine donor versus recipient. For explanation of the use of **double slash** (//) in chimeras, see section 4.1.

nuc ish(DXZ1×2)//[400]
  400 cells all representing the recipient.

//nuc ish(DXZ1,DYZ3)×1[400]
  400 cells all representing the donor.

nuc ish(DXZ1×2)[50]//(DXZ1,DYZ3)×1[350]

Fifty recipient XX cells and 350 donor XY cells were found using X and Y centromere probes.

nuc ish(DXZ1×2)[50]//(DXZ1,DYZ3)×1[300]/(DXZ1×1)[10]

Fifty recipient XX cells were found among 300 donor XY cells and loss of the Y chromosome from 10 cells, listed as though they are presumed from the donor.

nuc ish(ATM,D12Z3,D13S319,LAMP1,TP53)×2[200]

Normal hybridization pattern showing two copies of each of the probes used.

When normal and abnormal cells are found, the number of abnormal cells is listed over the total number of cells scored for each abnormal locus. The normal cells are not listed as it is implied that they are the remainder of the total. Probe sets co-hybridized are included in the same parentheses.

nuc ish(ATM×1,TP53×2)[200/400],(D12Z3×3,D13S319×2,LAMP1×2)[100/400]

Two separate hybridizations were performed. In the first, loss of *ATM* signal is found in 200 cells. The remaining cells scored, 200, had a normal signal pattern. In the second hybridization, a gain of signal is seen for D12Z3 in 100 cells. Three hundred cells show the normal pattern. Note, D13S319, *TP53* and *LAMP1* each showed a normal hybridization pattern in 400 cells analyzed.

nuc ish(D13S319×0)[100/400]

Homozygous deletion of D13S319 in 100 among 400 cells scored. Three hundred cells show a normal pattern.

nuc ish(D13S319×0)[100/400]/(D13S319×1)[50/400]

Homozygous deletion of D13S319 in 100 among 400 cells scored. Fifty cells show a heterozygous deletion. The remainder, 250 cells, show a normal pattern.

nuc ish(D13S319×0,LAMP1×2)[100/400]/(D13S319×1,LAMP1×2)[50/400]

Homozygous deletion of D13S319 in 100 among 400 cells scored. Fifty cells show a heterozygous deletion. The remainder, 250 cells, show a normal pattern. *LAMP1* is used as a control locus and shows two normal hybridization signals in the 400 interphase cells analyzed.

nuc ish(TP73×1,ANGPTL×2)[107/200],(ZNF443×2,GLTSCR×1)[105/200]

Interphase ish shows one *TP73* (maps to 1p36) signal with two *ANGPTL* signals (1q25) in 107 nuclei. In a second hybridization, interphase ish shows two *ZNF443* signals (19p13) with one *GLTSCR* signal (19q13) in 105 interphase nuclei. Thus, the specimen shows loss of both 1p and 19q.

nuc ish(D20Z1×2,D20S108×1)[100/200]

Interphase ish showing one copy of D20S108 as compared to two copies of the centromere probe in 100 cells.

13.4.2 Relative Position of Signals

If loci on two separate chromosomes are tested, they are expected under normal circumstances to be spatially separated and results are expressed:

nuc ish(ABL1,BCR)×2[400]

● = probe for *ABL1*
○ = probe for *BCR*

However, if they have become juxtaposed on one chromosome because of a t(9;22), the results are expressed with the first set of parentheses indicating the number of signals and the second set of parentheses describing the relative position of the signals to one another:

nuc ish(ABL1×2),(BCR×2),(ABL1 con BCR×1)[400]

● = probe for *ABL1*
○ = probe for *BCR*

An alternative way to express the results, more consistent with current FISH nomenclature:

nuc ish(ABL1,BCR)×2(ABL1 con BCR×1)[400]

If they have become juxtaposed on two chromosomes because of dual-fusion probes, the results are expressed:

nuc ish(ABL1×3),(BCR×3),(ABL1 con BCR×2)[400]
nuc ish(ABL1,BCR)×3(ABL1 con BCR×2)[400]

● = probe for *ABL1*
○ = probe for *BCR*

If a strange rearrangement has occurred resulting in the juxtaposition of one *BCR* locus with two *ABL1* loci the results are expressed:

nuc ish(ABL1×3),(BCR×3),(ABL1 con BCR×1),(ABL1 con BCR con ABL1×1)
[400]
nuc ish(ABL1,BCR)×3(ABL1 con BCR×1)(ABL1 con BCR con ABL1×1)[400]

● = probe for *ABL1*
○ = probe for *BCR*

Other possibilities using a dual-fusion probe are as follows:

nuc ish(ABL1×2),(BCR×2),(ABL1 con BCR×1)[400]
nuc ish(ABL1,BCR)×2(ABL1 con BCR×1)[400]
    Deletion of the *ABL1/BCR* fusion on one derivative chromosome.

nuc ish(ABL1×2),(BCR×3),(ABL1 con BCR×1)[400]
nuc ish(ABL1×2,BCR×3)(ABL1 con BCR×1)[400]
    Deletion of the *ABL1* locus from one fusion on one derivative chromosome.

nuc ish(ABL1×4),(BCR×4),(ABL1 con BCR×3)[198]
nuc ish(ABL1,BCR)×4(ABL1 con BCR×3)[198]
    Addition of one *BCR/ABL1* fusion through gain of one derivative chromosome.

If hybridization signals are normally juxtaposed because of close physical association of the respective loci on the same chromosome, normal results would be expressed as:

nuc ish(KAL1,STS)×2

● = probe for *KAL1*
○ = probe for *STS*

However, if the loci (example above) are separated because of a structural rearrangement of one X chromosome, the result is expressed:

nuc ish(KAL1×2),(STS×2),(KAL1 sep STS×1)
nuc ish(KAL1,STS)×2(KAL1 sep STS×1)

● = probe for *KAL1*
○ = probe for *STS*

13.4.2.1 Single Fusion Probes

nuc ish(ABL1,BCR)×2(ABL1 con BCR×1)[400]

Single fusion of the *ABL1* and *BCR* loci on a single chromosome identified through interphase analysis.

13.4.2.2 Single Fusion with Extra Signal Probes

nuc ish(ETV6×2,RUNX1×3)(ETV6 con RUNX1×1)[300/400]

ETV6 and *RUNX1* fusion with an extra *RUNX1* signal.

nuc ish(ETV6×1,RUNX1×3)(ETV6 con RUNX1×1)[300/400]

ETV6 and *RUNX1* fusion and deletion of *ETV6* with an extra *RUNX1* signal.

13.4.2.3 Dual Fusion Probes

nuc ish(ABL1,BCR)×3(ABL1 con BCR×2)[400]

Dual fusion of the *ABL1* and *BCR* loci in interphase.

nuc ish(D8Z1×2,MYC×3,IGH×3)(MYC con IGH×2)[100/200],
(IGH×3,BCL2×2)[100/200]

Dual fusion of the *MYC* and *IGH* loci in interphase in 100 cells. In a separate hybridization, three copies of *IGH* were seen as compared to two copies of *BCL2* in 100 cells, indicating an *IGH* translocation is present but it does not involve *BCL2*.

13.4.2.4 Break-Apart Probes

Given that break-apart probes are made of two probes, the short form does not convey that the normal situation is the presence of two fusion signals.

nuc ish(MLL×2)[400]
nuc ish(5′MLL,3′MLL)×2(5′MLL con 3′MLL×2)[400]

Two *MLL* fusion signals in interphase cells.

nuc ish(CBFB×2)[400]
nuc ish(5′CBFB,3′CBFB)×2(5′CBFB con 3′CBFB×2)[400]

Two *CBFB* fusion signals in normal interphase cells.

Abnormal cells show the separation of signals.

nuc ish(MLL×2)(5′MLL sep 3′MLL×1)[200]

Two *MLL* signals, but one has separated into the 5′ probe and the 3′ probe, presumably because of a translocation.

nuc ish(5′MLL×2,3′MLL×1)(5′MLL con 3′MLL×1)[200]

Two 5′*MLL* signals and one 3′*MLL* signal, presumably because of a deletion.

nuc ish(CBFB×2)(5′CBFB sep 3′CBFB×1)[200]

Two *CBFB* signals, but one has separated into the 5′ probe and 3′ probe, presumably because of an inversion or translocation.

nuc ish(5′IGH×3,3′IGH×2)(5′IGH con 3′IGH×1)[210/237]

Using the *IGH* break-apart probe, an *IGH* rearrangement was observed with an extra 5′ signal.

13.5   In situ Hybridization on Extended Chromatin/DNA Fibers (fib ish)

Hybridization can be carried out on extended chromatin/DNA fibers usually obtained from interphase nuclei, abbreviated **fib ish**. In this situation, the object of interest is the relative position of the loci at a particular chromosomal location. Where the order of the loci tested is known, they are recorded in the order pter to qter and the chromosomal band is indicated.

▬▬   = D15S11
●●●●   = *SNRPN*
xxxx   = *GABRB3*

fib ish 15q11.2(D15S11+,SNRPN+,GABRB3+)

Signifies that the three loci are present and in the order D15S11, *SNRPN*, *GABRB3*.

pter ——————▬▬▬—————●●●●——xxxx————————— qter

13.6   Reverse in situ Hybridization (rev ish)

Reverse in situ hybridization (**rev ish**) refers to the in situ hybridization of complex DNA probes derived from a test tissue to normal reference chromosomes. Chromosomes or chromosome segments with *enhanced* (**enh**) or *diminished* (**dim**) fluorescence intensity ratios indicate a relative increase or decrease of the copy number with regard to a basic euploid state. For example, a chromosome present in three copies in a near-diploid cell line would show an enhanced fluorescence intensity ratio, while a chromosome present in three copies in a near-tetraploid cell line would show a diminished ratio. This method can only reveal alterations in copy number of chromosomes or chromosomal segments.

Another method of reverse in situ hybridization uses DNA probes derived from parts of the genome from a test tissue, such as the DNA of sorted or microdissected marker chromosomes. In situ hybridization of such DNA probes to normal reference chromosomes or to DNA arrays reveals the composition of the isolated chromosome. This method is applicable both to constitutional and acquired abnormalities, and can reveal structural rearrangements not involving copy-number changes (inversions, balanced translocations).

13.6.1 Chromosome Analyses Using Probes Derived from Sorted or Microdissected Chromosomes

47,XX,+mar.rev ish 15q

Signifies that the marker chromosome is composed largely or wholly of material from 15q.

46,XY,add(5)(p15).rev ish der(5)t(5;10)(p15;q22)

Signifies a derivative chromosome consisting of part of the long arm of chromosome 10 translocated onto the short arm of chromosome 5.

13.7 Chromosome Comparative Genomic Hybridization (cgh)

Comparative genomic hybridization (**cgh**) involves the simultaneous hybridization of test DNA from cells suspected of carrying imbalance of chromosomes or chromosome segments with reference (control) DNA from cells with a known, usually normal, karyotype. The two DNA specimens are differentially labeled and hybridized to metaphase chromosomes from a karyotypically normal reference. CGH is a method which can detect relative copy-number changes. Only unbalanced gains or losses can be detected with this method. After chromosomal CGH, the karyotype can be re-written based on the knowledge gained from the FISH results. Alterations based on CGH alone can be written as follows:

47,XX,+mar.ish cgh enh(10)(p)

Signifies that the marker chromosome is composed mostly or wholly of material from 10p.

46,XX,add(7)(q36).ish cgh der(7)t(7;21)(q36;q22)enh(21)(q22)

Signifies that the additional material on 7q is composed of material from 21q22.

47,XY,+r.ish cgh enh(1)(q32q43),enh(12)(q13)

Signifies that the ring chromosome is composed of material from 1q32q43 and 12q13.

ish cgh enh(21)

Signifies extra copies of chromosome 21.

ish cgh dim(7)(q21qter)

Signifies a decrease in signal intensity from 7q21 to qter after chromosomal CGH.

ish cgh dim(18)(q21qter)

Signifies a decrease in signal intensity from 18q21 to 18qter.

ish cgh amp(1)(q31q32),enh(7),amp(7)(p12,q21),dim(10),enh(19)(p)

Signifies that the tissue examined had extra copies of chromosome 7 and 19p, a reduced number of chromosome 10, and amplified copy numbers of chromosome segment 1q31q32 and bands 7p12 and 7q21.

13.8   Multi-Color Chromosome Painting

24-color karyotyping and FISH banding are techniques used to paint chromosomes with a distinct color or spectrum of colors. They can be used as a tool to clarify the G-banded analysis. The karyotype can be re-written based on the knowledge gained from the FISH results using these techniques. The use of these FISH techniques should be stated in the report. No special nomenclature has been devised for these techniques. However, a nomenclature similar to that used for **wcp** (Section 13.3) may be used.

13.9   Partial Chromosome Paints

Microdissected segments of chromosomes can be used as *partial chromosome paints* (**pcp**). The nomenclature is similar to that of **wcp** (13.3), e.g.

46,XX,?dup(18)(p11.2p11.3).ish dup(18)(pcp18p+)

A questionable duplication on 18p is shown to be 18p material by a partial chromosome paint.

46,XY,inv(8)(p21q13).ish inv(8)(pcp8p++)

An inversion of chromosome 8 is confirmed by a partial chromosome paint.

14 Copy Number Detection

14.1 Microarray and Multiple Ligation-Dependent Probe Amplification (MLPA)

Microarray-based chromosome analysis has become an adjunct to traditional chromosome analysis and FISH and may become the laboratory method of choice for identifying chromosome abnormalities in the future. Microarrays can be constructed in two ways: with the use of large pieces of cloned DNA such as bacterial artificial chromosomes (BACs), or with small, synthetic sequences of DNA, termed oligonucleotides. Each BAC or oligonucleotide has a known position within the human genome. These DNA segments are spotted onto a solid support, usually a glass slide and serve as a target for the genomic DNA sample. In array-based comparative genomic hybridization (aCGH), this method uses a test DNA and a reference (control) DNA, differentially labeled and simultaneously applied to the microarray. In other non-CGH applications of microarray analysis, the patient DNA is hybridized to the microarray and compared by computer analysis to a pool of normal individuals. In either approach, the DNA in the patient is compared to the control or reference DNA and gains or losses can be detected. The nomenclature has been refined to encompass either type of array, built out of BACs or oligonucleotides. In this improved nomenclature, the number and type of clones used as targets (BAC, cosmid, fosmid, oligonucleotide, etc.) are not included. Two systems have been devised; a detailed description that includes the abnormal nucleotides as well as the bordering normal nucleotides, and a short description that includes only the abnormal nucleotides. At the discretion of the laboratory director or investigator, the clone name or accession number, gene name, GDB D-number, type of cloned DNA, and specified genome build [e.g. NCBI Build 35 (B35)] can be placed in the descriptive narrative or interpretation of the report.

Multiple ligation-dependent probe amplification (mlpa) is a relatively simple way of quantifying copy number at multiple loci at the same time. MLPA kits are commercially available for a wide range of specific applications. MLPA probes can also be synthesised by user laboratories and used for specific applications including the confirmation of copy number variation detected using aCGH. Like aCGH, MLPA provides only relative copy number with no information about position.

14.2 List of Symbols and Abbreviations

A list of symbols and abbreviations is also found in Chapter 3.

×   Multiplication sign, precedes the number of copies of a chromosomal region

~   Approximate sign, falls between the number of copies of a chromosomal region when the exact number cannot be determined

arr	Microarray
htz	Heterozygous, heterozygosity
hmz	Homozygous, homozygosity; used when one or two copies of a genome are detected, but previous, known heterozygosity has been reduced to homozygosity through a variety of mechanisms, e.g. loss of heterozygosity (LOH)
mlpa	Multiple ligation-dependent probe amplification (MLPA)

14.3   Examples of Microarray Nomenclature

If the results are normal using any type of array that has probes targeted to multiple loci across all chromosomes, the results are expressed as follows. The sex chromosomes are separated from the autosomes, which are listed first.

arr(1–22,X)×2   normal female
arr(1–22)×2,(XY)×1   normal male

The descriptive narrative, or interpretation, in the report should indicate the platform used, the resolution, and whether the array represents the entire genome of all chromosomes.

   If the results are *normal* using an array with probes restricted to a particular chromosome or chromosomal region, the results are expressed as follows. Note that there are no spaces between **arr** and the first parenthesis.

arr(2)×2

   Microarray analysis performed with clones specific for chromosome 2 shows a normal DNA copy number of two.

arr(X)×2

   Microarray analysis performed with clones specific for the X chromosome shows a normal DNA copy number of two in a female.

arr(2,6)×2

   Microarray performed with clones for chromosome 2 and chromosome 6 shows a normal DNA copy number of two.

   If the results are *abnormal*, list only the aberrations. Sex chromosome abnormalities should be listed first, followed by the lowest chromosome number, regardless of whether it is a copy number gain or loss. Only the band designations of the abnormal clones are shown. Similar to the **ish** nomenclature, which lists the aberrations from pter to qter on each chromosome arm, the aberrant nucleotides are listed from pter to qter, which is also consistent with the public databases derived from the Human Genome Project. Multiple nucleotides may be listed, separated by commas, or a dash may be used to indicate that the gain or loss encompasses the segment between the listed clones. Note that there is a space between the **arr** and the first abnormal chromosome number.

arr 4q32.2q35.1(163,146,681–183,022,312)×1
arr 4q32.2q35.1(163,002,425×2,163,146,681–183,022,312×1,184,322,231×2)

> Microarray analysis shows a loss of the long arm of chromosome 4 at bands q32.2 through q35.1, which is 19.8 Mb in size. The detailed description shows that the next neighboring proximal nucleotide that does not show a loss is 144,256 nucleotides away and the next neighboring distal nucleotide that does not show a loss is 1.3 Mb away from the alteration.

Because microarray analysis can only demonstrate a relative gain or loss of DNA, FISH analysis or karyotype confirmation or visualization is necessary to definitively demonstrate deletions, duplications, insertions, unbalanced translocations, etc.

The parental origin of the abnormality may follow the copy number (×1, ×3, etc.). There is a space between the copy number and the inheritance abbreviation (dn, mat, pat), but no space if the inheritance abbreviation follows a parenthesis in the detailed system.

arr 4q32.2q35.1(163,146,681–183,022,312)×1 dn
arr 4q32.2q35.1(163,002,425×2,163,146,681–183,022,312×1,184,322,231×2)dn

arr Xq25(126,228,413–126,535,347)×0 mat
arr Xq25(126,023,321×1,126,228,413–126,535,347×0,126,556,900×1)mat

> Microarray analysis shows a loss of the long arm of the X chromosome at band q25 in a male. The hemizygous loss is 306,934 nucleotides. The next neighboring proximal nucleotide that does not show a loss is 205,092 nucleotides away and the next neighboring distal nucleotide that does not show a loss is 21,553 nucleotides away from the alteration. This deletion was inherited from the mother.

arr Xq25(126,228,413–126,535,347)×1
arr Xq25(126,023,321×2,126,228,413–126,535,347×1,126,556,900×2)

> Same abnormality as the above example, but found in a female.

arr Xp22.31(6,467,202–8,091,950)×0 mat

> Microarray analysis in a male shows a loss of the short arm of the X chromosome at band p22.31, inherited from a carrier mother.

arr Xp11.22(53,215,290–53,986,534)×2 mat

> Microarray analysis in a male shows a gain of the short arm of the X chromosome at band p11.22, inherited from a carrier mother.

arr Xp11.22(53,215,290–53,986,534)×3

> Same abnormality as the above example, but found in a female.

arr 11p12(37,741,458–39,209,912)×3
arr 11p12(37,003,221×2,37,741,458–39,209,912×3,39,752,007×2)

> Microarray analysis shows a single copy gain of the short arm of chromosome 11 at band p12. The duplication is ~1.47 Mb in size. The next neighboring distal clone that does not show a gain is 738,237 away from the alteration and the next neighboring proximal clone that does not show a gain is 542 kb away from the alteration.

arr 17p11.2(16,512,256–20,405,113)×3 dn

> Microarray analysis shows a single copy gain of the short arm of chromosome 17 at band p11.2. The duplication is ~3.9 Mb in size and is de novo in origin.

Observations combined with banded chromosome analysis and microarrays can be expressed by using the symbol **arr** followed by a space and the chromosome region, band or sub-band designation of the locus. A period (**.**) precedes the microarray nomenclature.

47,XY,+mar.arr 1p13.1p11.2(117,596,421–121,013,236)×3 dn

> Microarray analysis shows a single copy gain of the short arm of chromosome 1, spanning ~3.4 Mb in size, likely identifying the marker chromosome. Because most microarrays will not contain the heterochromatin near the centromeres, the centromeric bands are rarely included in the nomenclature of rings and markers after microarray analysis, although the centromere is probably included in the aberration and would need to be confirmed by FISH. An amended result after FISH analysis could be written as:

47,XY,+mar.ish r(1)(p13.1q11?)(D1Z1+).arr 1p13.1p11.2(117,596,421–121,013,236)×3 dn

47,XY,+mar.arr 1p13.1p12(117,596,421–121,013,236)×3,15q25.1q26.3(78,932,946–100,201,136)×3 dn

> Microarray analysis shows a single copy gain of the short arm of chromosome 1 and a single copy gain of the long arm of chromosome 15, likely identifying a complex marker comprised of material from chromosome 1 and chromosome 15. The **dn** at the end of the string indicates that the whole string is de novo.

46,XX.arr 3p12.3p12.2(80,395,073–83,498,191)×3 mat,12p12.1(23,543,231–23,699,047)×1 dn

> Normal female chromosome analysis showing a gain of the 3p by microarray analysis, inherited from the mother, and a loss of 12p at band 12p12.1 of de novo origin.

arr 8q22.3q24.3(105,171,556–146,201,911)×3,15q26.2q26.3(96,062,102–100,201,136)×1

> Microarray analysis shows a single copy gain of 8q and a single copy loss of 15q. Double segmental imbalances are indicative of unbalanced translocations. However, microarrays can only detect relative imbalances in DNA copy number. After chromosome analysis and FISH visualization, the nomenclature can be written as follows:

46,XY,der(15)t(8;15)(q22.3;q26.2)mat.ish der(15)t(8;15)(RP11-1143I12+,RP11-14C10–).arr 8q22.3q24.3(105,171,556–146,201,911)×3,15q26.2q26.3(96,062,102–100,201,136)×1

> **mat** is placed after the conventional cytogenetic result because the derivative was determined to be inherited from a balanced translocation in the mother.

Note that the conventional cytogenetic banding assignments are those derived from banded chromosomes, while the array banding assignments are those derived from genome browsers. These are not always concordant with each other.

arr 20q13.2q13.33(51,001,876–62,375,085)×1,22q13.33(48,533,211–49,525,263)×3

> Microarray analysis shows a single copy loss of 20q and a single copy gain of 22q.

arr 4q28.1q28.2(128,184,801–129,319,376)×3 mat,16p11.2(29,581,254–
30,066,186)×3 pat

> Microarray analysis shows a gain of 4q, inherited from the mother, and a gain of 16p, inherited
> from the father.

arr 9p24.3(1,310,386–1,709,409)×1 mat,9p22.3p22.2(16,455,330–16,763,471)×1 dn,
18q22.1q22.3(62,747,805–67,920,791)×1 dn

> Microarray analysis shows three abnormalities. The first abnormality is a loss of maternal origin; the
> other two abnormalities are de novo losses. Therefore, the inheritance of each is listed after the spe-
> cific gain or loss. Note that the two abnormalities on chromosome 9 are listed from pter to qter.

arr 14q31.2(82,695,844–82,855,387)×1,14q32.22(105,643,093–106,109,395)×3

> Microarray analysis shows two abnormalities on chromosome 14. Note that the abnormalities are
> shown from pter to qter, irrespective of whether they are gains or losses.

arr 9p24.3p13.1(204,166–38,756,057)×1,18q22.1(63,877,984–64,683,663)×1,
21q11.2q21.1(13,600,026–20,175,986)×3

> Microarray analysis shows three abnormalities; a deletion of the short arm of the 9p covered by
> the array, a deletion of the long arm of chromosome 18 and a duplication of the distal long arm of
> chromosome 21. Note that the chromosomes are listed in numerical order, regardless of whether
> they show a gain or loss.

ish del(18)(q22.1q22.1)(RP11-106E15–),der(9)t(9;21)(p11;q21.1)(RP11-59O6–,
RP11-78J18+).arr 9p24.3p13.1(204,166–38,756,057)×1,18q22.1(63,877,984–
64,683,663)×1,21q11.2q21.1(13,600,026–20,175,986)×3

> Same example as above, showing the FISH nomenclature. Note that the FISH nomenclature fol-
> lows the rules in Chapter 13, resulting in the deletion 18 listed first, followed by the derivative
> chromosome 21. However, the microarray nomenclature only shows the relative gains and losses
> of DNA segments, which are listed in numerical order, regardless of the structural changes of the
> chromosomes.

arr 1p36.33p36.32(827,048–3,736,354)×3,1q41q44(221,649,655–247,175,095)×1

> Microarray analysis shows a gain of the short arm of chromosome 1 and a loss of the long arm of
> chromosome 1. This result may indicate a duplication/deletion recombinant chromosome from
> an inversion parent, but further studies of the parents and/or child by FISH or chromosome anal-
> ysis is required.

arr 18p11.32(102,328–2,326,882)×1,18q21.32q23(56,296,522–76,093,443)×3 pat

> Microarray analysis shows a loss of the short arm of chromosome 18 and a gain of the long arm of
> chromosome 18. Chromosome analysis in the father demonstrated a balanced pericentric inversion.
> Thus, this is a duplication/deletion recombinant chromosome from an inversion carrier parent.

46,XY,rec(18)dup(18q)inv(18)(p11.32q21)pat.arr 18p11.32(102,328–2,326,882)×1,
18q21.32q23(56,296,522–76,093,443)×3

> Same example as above, showing conventional cytogenetic nomenclature as part of the string.

arr Yq11.23(26,887,746–27,019,505)×0,20q13.2q13.33(51,840,606–62,375,085)×3

Microarray analysis shows a loss of the long arm of the Y chromosome and a gain of the long arm of chromosome 20. Note that the sex chromosome abnormality is listed first.

46,X,der(Y)t(Y;20)(q11.23;q13.2).arr Yq11.23(26,887,746–27,019,505)×0, 20q13.2q13.33(51,840,606–62,375,085)×3

Same example as above, showing the banded chromosome nomenclature. Note that there is no normal Y chromosome in this individual.

46,XY,der(20)t(Y;20)(q11.23;q13.2).arr Yq11.23(26,887,746–27,019,505)×2, 20q13.2q13.33(51,840,606–62,375,085)×1

Microarray analysis shows an unbalanced translocation, derived from rearrangement between the long arm of the Y chromosome, resulting in a gain of distal Yq, and the long arm of one chromosome 20, resulting in a deletion of distal 20q. Note that the array nomenclature lists the sex chromosome abnormality first and that there is a normal Y chromosome, in addition to the derivative chromosome 20, in this individual.

46,XX.arr Xp22.3(6,923,924–7,253,485)×3,5q14.3(88,018,766–89,063,989)×1

Microarray analysis shows a single copy gain of the short arm of the X chromosome and a single copy loss of the long arm of chromosome 5. Note that the sex chromosome abnormality is listed first.

46,X,der(Y)t(X;Y)(p22.33;q12).arr Xp22.33(701–2,679,502)×3,Xp22.33p22.2 (2,709,521–15,955,588)×2,Yq11.22q11.23(16,139,805–27,177,529)×0

Chromosome and microarray analysis shows a single copy gain of the X chromosome from two regions of Xp and loss of the long arm of the Y chromosome, resulting from an unbalanced translocation between the short arm of the X chromosome and the long arm of the Y. The two regions of Xp are shown separately because the gain of the pseudoautosomal region results in three total copies and the gain proximal to the pseudoautosomal region results in two total copies.

ish mos del(2)(q11.2q13)(RP11-11P22–)[10/30].arr 2q11.2q13(90,982,729– 112,106,760)×1~2

FISH and microarray analyses show a mosaic deletion in the long arm of chromosome 2. The approximate sign (~) is used to indicate that the number of copies varies from 1 to 2 copies.

47,XY,+mar.ish der(2)(q11.2q13)(D2Z2+)[5/30].arr 2q11.2q13(90,982,729– 112,106,760)×2~3

FISH and microarray analysis demonstrate a mosaic marker chromosome, derived from chromosome 2. The approximate sign (~) is used to indicate that the number of copies of this region varies from 2 to 3.

47,XX,+mar.arr 21q11.2q21.1(13,461,349–17,308,947)×4,21q22.3(46,222,759– 46,914,885)×3

Microarray analysis shows a two copy gain of 21q11.2q21.1 and a single copy gain of 21q22.3, indicating that the marker chromosome is likely a complex rearrangement involving two different segments of chromosome 21, resulting in partial tetrasomy of proximal 21q and partial trisomy of distal 21q.

arr 15q11.2q13.3(20,366,669–30,226,235)×4

Microarray analysis shows a two copy gain of proximal 15q, resulting in tetrasomy 15q11.2q13.3. Distinguishing a marker chromosome from a triplication of this region requires FISH.

arr 12p13.33p11.1(84,917–34,382,567)×2~4

Microarray analysis shows a two copy gain of the short arm of chromosome 12, resulting in tetrasomy 12p. Although this result likely indicates an isochromosome of 12p, such as those found in Pallister Killian syndrome, FISH or chromosome analysis is required to confirm. The approximate sign is used to indicate that the number of copies of this region varies from 2 to 4.

arr 18p11.32q23(102,328–76,093,443)×3

Microarray analysis shows a single copy gain of the entire chromosome 18, likely indicating trisomy 18.

arr 21q11.2q22.3(9,931,865–46,914,745)×3

Microarray analysis shows a single copy gain of the entire long arm of chromosome 21, likely indicating trisomy 21. Note that most microarrays will not have coverage of the repetitive short arm sequences; thus the short arm is not designated. Trisomy 21 is implied, but FISH or cytogenetic confirmation is indicated to exclude a Robertsonian or other translocation.

arr 18p11.32p11.21(102,328–15,079,388)×1,18q22.3q23(69,172,132–76,093,443)×1

Microarray analysis shows a single copy loss of the distal short arm of chromosome 18 and a single copy loss of the distal long arm of chromosome 18, likely indicating a ring chromosome 18, although FISH or chromosome analysis is required to confirm.

Single nucleotide polymorphisms (SNP) can be identified using certain types of oligonucleotide microarrays. The use of SNP arrays may uncover regions of homozygosity that have been reduced from previously, known heterozygosity. The abbreviations **htz** and **hmz** can be used to define the zygosity of the chromosomal region.

arr 11p12(37,741,458–39,209,912)×2 hmz

SNP array analysis shows homozygosity in the short arm of chromosome 11, at band p12, ~1.47 Mb in size.

arr 11p12(37,741,458–39,209,912)×2 hmz mat

SNP array analysis shows homozygosity in the short arm of chromosome 11, at band p12, ~1.47 Mb in size and of maternal origin.

upd(16)mat.arr 16p13.3q23.1(31,010–78,001,824)×2 htz,16q23.1q24.3 (78,003,664–88,690,776)×2 hmz

SNP array analysis shows heterozygosity for the majority of chromosome 16 in the region 16p13.3q23.1 and homozygosity for the distal long arm of chromosome 16 in the region 16q23.1q24.3. In combination with subsequent SNP array data obtained from the parents, uniparental disomy for a maternally derived chromosome 16 was identified.

arr 15q11.2q26.3(18,427,103–100,338,915)×2 hmz pat,21q11.21q22.3(9,887,804–46,944,323)×2 hmz pat

SNP array analysis shows homozygosity for the entire long arms of chromosomes 15 and 21, respectively. Based on additional SNP array analyses in the parents, both regions of homozygosity reveal uniparental isodisomy obtained from the father.

Copy Number Detection

14.4   Examples of MLPA Nomenclature

46,XY.mlpa (P023)×2

Normal male karyotype and normal copy number of all probes within the P023 kit.

46,XX.mlpa 22q11.2(P023)×1

Normal female karyotype with a deletion of probes mapping to the DiGeorge critical region in 22q11.2. All other probes at normal copy number.

46,XY.mlpa (P070)×2

Normal male karyotype and normal copy number of all probes within the P070 subtelomeric probe kit.

46,XX.mlpa 1psubtel(P070)×1

Normal female karyotype with a subtelomeric deletion of 1p.

mlpa X(P095)×2,13,18,21(P095)×2

Normal female with two copies of all targets for the X and chromosomes 13, 18, 21 within the P095 kit.

mlpa X,Y(P095)×1,13,18,21(P095)×2

Normal male with one copy of all targets for chromosomes X and Y and two copies of all targets for chromosomes 13,18 and 21 within the P095 kit.

mlpa X,Y(P095)×1,13(P095)×3,18,21(P095)×2

Male with three copies of all targets on chromosome 13, two copies of all targets for chromosomes 18 and 21, and one copy of all targets for the X and Y chromosomes within the P095 kit.

mlpa X(P095)×1,13,18,21(P095)×2

Female with one copy of all targets for the X chromosome within the P095 kit and two copies of all targets for chromosomes 13,18 and 21.

arr 8p23.1(8,479,797–11,897,580)×1.mlpa 8p23.1(P139)×1

Microarray analysis shows a deletion of 8p at band 8p23.1 which was confirmed using MLPA with the P139 kit.

arr 8p23.1(8,479,797–11,897,580)×1.mlpa (GATA4)×1

Same example as above confirmed using a MLPA probe targeted at the *GATA4* gene within the deleted interval.

15 References

Caspersson T, Farber S, Foley GE, Kudynoski J, Modest EJ, Simonsson E, Wagh U, Zech L: Chemical differentiation along metaphase chromosomes. Exp Cell Res 49:219–222 (1968).

Caspersson T, Lomakka G, Zech L: The 24 fluorescence patterns of human metaphase chromosomes – distinguishing characters and variability. Hereditas 67:89–102 (1972).

Chicago Conference (1966): Standardization in Human Cytogenetics. Birth Defects: Original Article Series, Vol 2, No 2 (The National Foundation, New York 1966).

Cremer T, Landegent J, Bruckner A, Scholl HP, Schardin M, Hager HD, Devilee P, Pearson P, van der Ploeg M: Detection of chromosome aberrations in the human interphase nucleus by visualization of specific target DNAs with radioactive and non-radioactive in situ hybridization techniques: diagnosis of trisomy 18 with probe L1.84. Hum Genet 74:346–352 (1986).

Denver Conference (1960): A proposed standard system of nomenclature of human mitotic chromosomes. Lancet i:1063–1065 (1960).

Dutrillaux B: Obtention simultanée de plusieurs marquages chromosomiques sur les mêmes préparations, après traitement par le BrdU. Humangenetik 30:297–306 (1975).

Ford CE, Hamerton JL: The chromosomes of man. Nature 178:1020–1023 (1956).

Francke U: High-resolution ideograms of trypsin-Giemsa banded human chromosomes. Cytogenet Cell Genet 31:24–32 (1981).

Francke U: Digitized and differentially shaded human chromosome ideograms for genomic applications. Cytogenet Cell Genet 65:206–218 (1994).

Francke U, Oliver N: Quantitative analysis of high-resolution trypsin-Giemsa bands on human prometaphase chromosomes. Hum Genet 45:137–165 (1978).

Guan XY, Meltzer PS, Trent JM: Rapid generation of whole chromosome painting probes (WCPs) by chromosome microdissection. Genomics 22:101–107 (1994).

ISCN (1978): An International System for Human Cytogenetic Nomenclature. Birth Defects: Original Article Series, Vol 14, No 8 (The National Foundation, New York 1978); also in Cytogenet Cell Genet 21:309–404 (1978).

ISCN (1981): An International System for Human Cytogenetic Nomenclature – High Resolution Banding. Birth Defects: Original Article Series, Vol 17, No 5 (March of Dimes Birth Defects Foundation, New York 1981); also in Cytogenet Cell Genet 31:1–23 (1981).

ISCN (1985): An International System for Human Cytogenetic Nomenclature, Harnden DG, Klinger HP (eds), Birth Defects: Original Article Series, Vol 21, No 1 (March of Dimes Birth Defects Foundation, New York 1985).

ISCN (1991): Guidelines for Cancer Cytogenetics, Supplement to An International System for Human Cytogenetic Nomenclature, Mitelman F (ed), (S Karger, Basel 1991).

ISCN (1995): An International System for Human Cytogenetic Nomenclature, Mitelman F (ed), (S Karger, Basel 1995).

ISCN (2005): An International System for Human Cytogenetic Nomenclature, Shaffer LG, Tommerup N (eds), (S Karger, Basel 2005).

Jhanwar SC, Burns JP, Alonso ML, Hew W, Chaganti RSK: Mid-pachytene chromomere maps of human autosomes. Cytogenet Cell Genet 33:240–248 (1982).

Kallioniemi A, Kallioniemi OP, Sudar D, Rutovitz D, Gray JW, Waldman F, Pinkel D: Comparative genomic hybridization for molecular cytogenetic analysis of solid tumors. Science 258:818–821 (1992).

Landegent JE, Jansen in de Wal N, Dirks RW, Baao F, van der Ploeg M: Use of whole cosmid cloned genomic sequences for chromosomal localization by non-radioactive in situ hybridization. Hum Genet 77:366–370 (1987).

Lichter P, Cremer T, Borden J, Manuelidis L, Ward DC: Delineation of individual human chromosomes in metaphase and interphase cells by in situ suppression hybridization using recombinant DNA libraries. Hum Genet 80:224–234 (1988).

Lichter P, Tang CJ, Call K, Hermanson G, Evans GA, Housman D, Ward DC: High-resolution mapping of human chromosome 11 by in situ hybridization with cosmid clones. Science 247:64–69 (1990).

Liehr T, Claussen U, Starke H: Small supernumerary marker chromosomes (sSMC) in humans. Cytogenet Genome Res 107:55–67 (2004).

Liehr T, Starke H, Heller A, Kosyakova N, Mrasek K, Gross M, Karst C, Steinhaeuser U, Hunstig F, Fickelscher I, Kuechler A, Trifonov V, Romanenko SA, Weise A: Multicolor fluorescence in situ hybridization (FISH) applied to FISH-banding. Cytogenet Genome Res 114:240–244 (2006).

London Conference on the Normal Human Karyotype. Cytogenetics 2:264–268 (1963).

Magenis RE, Barton SJ: Delineation of human prometaphase paracentromeric regions using sequential GTG- and C-banding. Cytogenet Cell Genet 45:132–140 (1987).

Paris Conference (1971): Standardization in Human Cytogenetics. Birth Defects: Original Article Series, Vol 8, No 7 (The National Foundation, New York 1972); also in Cytogenetics 11:313–362 (1972).

Paris Conference (1971), Supplement (1975): Standardization in Human Cytogenetics. Birth Defects: Original Article Series, Vol 11, No 9 (The National Foundation, New York 1975); also in Cytogenet Cell Genet 15:201–238 (1975).

Parra I, Windle B: High resolution visual mapping of stretched DNA by fluorescent hybridization. Nat Genet 5:17–21 (1993).

Patau K: The identification of individual chromosomes, especially in man. Am J Hum Genet 12:250–276 (1960).

Pinkel D, Landegent J, Collins C, Fuscoe J, Segraves R, Lucas J, Gray JW: Fluorescence in situ hybridization with human chromosome-specific libraries: detection of trisomy 21 and translocations of chromosome 4. Proc Natl Acad Sci USA 85:9138–9142 (1988).

Pinkel D, Segraves R, Sudar D, Clark S, Poole I, Kowbel D, Collins C, Kuo WL, Chen C, Zhai Y, Dairkee SH, Ljung BM, Gray JW, Albertson DG: High resolution analysis of DNA copy number variation using comparative genomic hybridization to microarrays. Nat Genet 20:207–211 (1998).

Tjio JH, Levan A: The chromosome number of man. Hereditas 42:1–16 (1956).

Trask BJ: Fluorescence in situ hybridization: applications in cytogenetics and gene mapping. Trends Genet 7:149–154 (1991).

Viegas-Pequignot E, Dutrillaux B: Une méthode simple pour obtenir des prophases et des prometaphases. Annls Genet 21:122–125 (1978).

Wiegant J, Kalle W, Mullenders L, Brookes S, Hoovers JM, Dauwerse JG, van Ommen GJ, Raap AK: High-resolution in situ hybridization using DNA halo preparations. Hum Mol Genet 1:587–591 (1992).

Wiegant J, Wiesmeijer CC, Hoovers JM, Schuuring E, d'Azzo A, Vrolijk J, Tanke HJ, Raap AK: Multiple and sensitive fluorescence in situ hybridization with rhodamine-, fluorescein-, and coumarin-labeled DNAs. Cytogenet Cell Genet 63:73–76 (1993).

Wyandt HE, Tonk VS (eds): Atlas of Human Chromosome Heteromorphisms (Springer, New York 2008).

Yunis JJ: High resolution of human chromosomes. Science 191:1268–1270 (1976).

Yunis JJ, Sawyer JR, Ball DW: The characterization of high-resolution G-banded chromosomes of man. Chromosoma 67:293–307 (1978).

16 Members of the ISCN Standing Committee and Consultants

Members of the Standing Committee

Lynda J. Campbell
Victorian Cancer Cytogenetics Service
St. Vincent's Hospital
Melbourne, Vic., Australia

Myriam Chaabouni
Faculté de Medecine Tunis
Laboratory of Human Genetics
Tunis, Tunisia

Yoshimitsu Fukushima
Department of Medical Genetics
Shinshu University School of Medicine
Nagano, Japan

Christine J. Harrison
Leukaemia Research Cytogenetics Group
Cancer Sciences Division
University of Southampton
Southampton, UK

Prochi Madon
Department of Assisted Reproduction and Genetics
Jaslok Hospital and Research Centre
Mumbai, India

Nils Mandahl
Department of Clinical Genetics
Lund University Hospital
Lund, Sweden

Kathleen W. Rao
Department of Pediatrics, Cytogenetics Laboratory
University of North Carolina
Chapel Hill, N.C., USA

Carla Rosenberg
Department of Genetics and Evolutionary Biology
Institute of Biosciences
University of Sao Paulo
Sao Paulo, Brazil

Albert A. Schinzel
Institute for Medical Genetics
University of Zurich
Schwerzenbach, Switzerland

Lisa G. Shaffer (Chair)
Signature Genomic Laboratories
Spokane, Wash., USA

Marilyn L. Slovak
Department of Cytogenetics
City of Hope National Medical Center
Duarte, Calif., USA

Consultants

John C. Barber National Genetics Reference Laboratory (Wessex)
Salisbury, UK

Thomas Liehr Institute of Human Genetics and Anthropology
Jena, Germany

Holger Tönnies Institute for Human Genetics
Christian-Albrechts University
Kiel, Germany

Acknowledgements

The 2008 conference and the ISCN (2009) were made possible by generous contributions from Karger and The Danish National Research Foundation. The Committee gratefully acknowledges Martina Guttenbach and Michael Schmid, University of Würzburg, Germany, for copy editing the manuscript; Aaron Theisen, Signature Genomic Laboratories, Spokane, Wash., USA, for his substantial assistance and skillful editing of the manuscript; Rhett P. Ketterling, Mayo Clinic, Rochester, Minn., for providing FISH examples, review, and discussion; Nicole Chia, Canberra Hospital, Canberra City, A.C.T., Australia, for creating the revised idiograms; and Ros Hastings, UK National External Quality Assessment Scheme (NEQAS) for Clinical Cytogenetics, Oxford, UK, and Nicole de Leeuw, Radboud University Medical Centre, Nijmegen, the Netherlands, for reviewing Chapter 14.

17 Appendix

Diagrammatic representation of human chromosome bands as observed with the Q-, G-, and R-staining methods; centromeric regions are representative of Q-staining method only (Paris Conference, 1971).

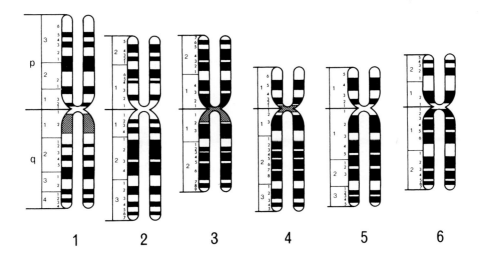

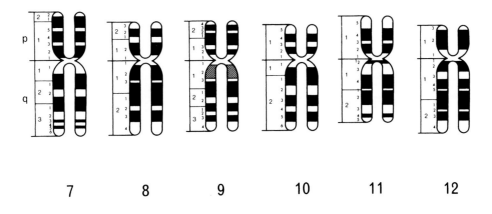

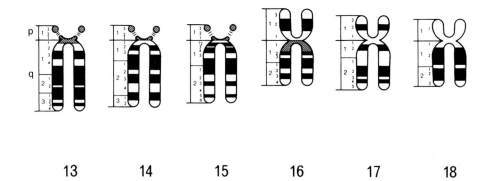

| 13 | 14 | 15 | 16 | 17 | 18 |

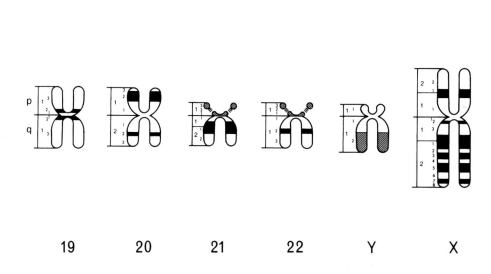

| 19 | 20 | 21 | 22 | Y | X |

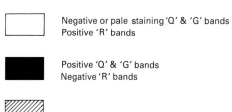

Negative or pale staining 'Q' & 'G' bands
Positive 'R' bands

Positive 'Q' & 'G' bands
Negative 'R' bands

Variable bands

ISCN 2009

18   Index

Pseudodicentric, -tricentric   68
Pseudodiploid, -triploid   95
Pulverization   38, 87

Q-band   7, 8, 13
Quadriradial   38, 85
Quadrivalent   97
Quadruplication   38, 75
Questionable identification   38, 49
Question mark   38, 40, 49, 60, 65

R-band   7, 10, 13, 34
Rearrangement, balanced   46, 47, 80
    complex   41, 44, 66, 79
    four-break   44, 79
    three-break   43, 78
    two-break   42, 77
    unbalanced   44, 80
Reciprocal translocation   46, 77, 86
Recombinant chromosome   38, 46, 62
Region, abnormally banded   70
    definition   9
    homogeneously staining   37, 70
Ring chromosome   38, 75
Robertsonian translocation   38, 81
Roman numeral   38, 97

Satellite   38, 53
Satellite stalk   38, 53
Semicolon   38, 39, 97, 106
Separated signal   106, 115
Sex, chromatin   8
    chromosome abnormality   56
Short system   42
Sideline   89
Sign, approximate   36, 40, 50
    equal   37, 98
    minus   37, 40, 53, 55, 97, 105
    multiplication   37, 40, 83, 106, 112, 121
    plus   38, 40, 53, 55, 97, 105
Signal, adjacent   106, 115
    amplified   106
    connected   106, 115
    intensity   106, 111, 118
    number   112
    position   115
    separated   106, 115

Single nucleotide polymorphism (SNP)   127
Sister chromatid exchange   38, 85
Slant line   38, 40, 88
Square brackets   36, 40, 89, 112
Stemline   38, 89
Sub-band   10
Subclone   89
Suppressed centromere   68
Symbols, list   36, 105, 121

T-band   7
Telomeric association   38, 77
Terminal deletion   61
Tetraploid   55, 94
Three-break rearrangement   43, 78
Translocation, balanced   80, 118
    complex   79
    jumping   82
    reciprocal   46, 77, 86
    Robertsonian   81
    segregation   46, 47
    unbalanced   80
    whole-arm   80
Tricentric chromosome   38, 76, 82
Triplication   82
Triploid   94
Triradial   38, 85
Trivalent   97
Two-break rearrangement   42, 77

Unbalanced rearrangement   44, 80
Uncertainty, breakpoint localization   50
    chromosome number   50
Underlining   38, 39, 66, 77
Uniparental disomy   38, 58
Univalent   97
Unknown material   60
Unrelated clones   93

Variable chromosome region   13, 38, 53

Whole-arm translocation   80, 82
Whole chromosome paint   106

X-chromatin   8

Y-chromatin   8